The Science of the Clouds

Other books by R. A. R. Tricker

Paths of the Planets
The Contribution of Science to Education
The Assessment of Scientific Speculation
The Science of Movement (with B. J. K. Tricker)
Bores, Breakers, Waves and Wakes
Introduction to Meteorological Optics

1. *Cirrus cloud showing the typical wind-swept, fibrous appearance. Cirrus clouds are composed of ice particles and the fibres are caused by the particles settling slowly into air travelling at different speeds to that of the air in which the cloud itself originated.*

2. *Cirrus nebulosus showing a well-developed halo. The sky is filled uniformly with a fog of ice crystals, the halo being formed by the bending of the light in ice prisms of angle 60°. The sky appears brighter outside the halo than it does inside.*

3. *Alto-cumulus cloud – the cloud of the familiar mackerel sky. Being high in the sky it receives the last rays of the setting sun and the first rays at sunrise, and can produce some very beautiful effects.*

The Science
of the Clouds

R. A. R. Tricker, M.A., Ph.D.

American Elsevier
Publishing Company Inc
New York

Mills & Boon
Publishers
London

ST. PHILIPS COLLEGE LIBRARY

First published 1970 by Mills and Boon Limited,
50 Grafton Way, Fitzroy Square, London W1A 1DR

First published 1970 U.S.A. by American Elsevier Publishing Co, Inc,
52 Vanderbilt Avenue, New York, New York 10017

© R. A. R. Tricker 1970

All rights reserved. No part of this publication may be
reproduced, stored in a retrieval system, or transmitted,
in any form or by any means, electronic, mechanical,
photocopying, recording or otherwise, without the prior
permission of the copyright owner.

I.S.B.N. 0 263 69982 X
(U.S.) 444 19656 0

Library of Congress Catalog Number: 74 116584

551.576
T823

\

Made and printed by offset in Great Britain by
William Clowes and Sons, Limited, London and Beccles

Contents

33848

List of Colour Plates

List of Black and White Plates

Preface

There are few branches of physics which do not have some impact on the science of the clouds and to write a complete book on the subject would involve dealing with nearly the whole of physics. Needless to say, this little book makes no attempt to do any such thing. What it does set out to do is to consider some of the most elementary aspects of the topic likely to be of interest to the amateur observer. Even a very little knowledge adds enormously to the interest which can be found in the sky and it can open a new area of natural history for those attracted to the observation of nature. For those also who are embarking on a study of physics the science of the clouds affords an attractive peg on which to hang a great deal of the early stages in the subject. For both categories of readers the interest can be heightened by practical work and the last chapter of the book contains suggestions for certain things which might be attempted – things to make, to think about, and to do. In this much has been left to the ingenuity of the reader. The mere following out of instructions with no practical problems to be solved and no difficulties overcome is, by comparison, a poor thing, as likely to add to tedium as to increase interest.

Most of the photographs illustrating the text were taken by the author. He wishes to record his thanks, however, to Miss G. Jones for the transparency from which Colour Plate 17 was reproduced, and to Mrs E. A. Holland for that used in Colour Plate 22.

R. A. R. Tricker

November 1969.

Chapter I

Classification

To any person who takes a delight in colour few objects are of greater beauty than the clouds. It is not only at times of sunset and sunrise that their colours are remarkable. They are worth looking at for this reason alone at all times. The gradations of tone which they show are more delicate than any to be found in the finest water colour, and they show the most exquisite hues. Many ascend to heights greater than that of the highest mountain peak, they are usually lit by the most brilliant light and they are there for all to see, for practically no effort. They are well worth looking at.

A little time spent in looking at the clouds will soon show that there are a number of different kinds which can be distinguished from one another. It is useful to give names to some of the most common and important of them, so that we can refer to them easily, and by using the names which are commonly given we can understand something of what is written about them. In actual fact there are very few names in common use, and although they are Latin ones for the most part, the task of learning them is nothing like as difficult as learning the Latin and Greek names in biology.

There are a number of ways in which clouds may differ and systems of naming them can be set up in different ways. Clouds may be named on account of the way in which they are thought to have been formed. As we must start without knowing anything about this, it will not be a very suitable method to adopt as a first step. Clouds may also be named on account of what they are made of. Again we do not know this to start with, and although the groups we want to distinguish are often made up differently, those to which we shall give names must be decided upon in another way. Height is another possibility on which clouds could be grouped. A good idea of the height of the different kinds of clouds had already been obtained long before aeroplanes, or even balloons, were able to fly up into them. We can measure the height of some of them ourselves, and the heights of some other kinds were known from the

ST. PHILIPS COLLEGE LIBRARY

various mountain peaks which were high enough to go up into them. Our names, however, can be given, in the first place, to groups which possess different shapes.

The name cirrus, which means fibrous, is given to any cloud which is made up of wisps of hair-like fibres. Such clouds have often a very wind-swept appearance, with the strands drawn out in one direction. These clouds are usually formed at great height, clear of the tops of the peaks of the Alps, for example, but they can occur at almost any height. Streams of such clouds can sometimes be seen in the wind blowing over quite low hills – such as Snaefell in the Isle of Man, which is only 770 metres (2,034 ft) high, and the clouds can be seen to be little higher than this. Haloes, or parts of haloes, are sometimes to be seen in these clouds, and the radius of the circle which they form – 22° – shows that they are made of ice crystals. Ice crystals have faces at angles which form a prism of angle 60°, and 22° is the angle through which such a prism would bend most of the light. We shall see later, in Chapter XI, p. 112, how, by spinning a 60° water prism in a beam of light, the angle of the halo may be forecast, approximately. In the Antarctic, where temperatures are often very much below freezing-point, ice particle clouds can form at ground level. The air becomes filled with a fog of ice dust. It is thought that the fibres of the cirrus cloud are produced as the ice dust settles very slowly out of the cloud in which it is formed, and is carried along by winds blowing at different speeds from that of the air in which the cloud itself lies. The cirrus cloud in Colour Plate 1 shows the fibres in a rather unusual direction. They seem to point up and down. This is because they are directed towards us and we, being a long way below them, see the particles being carried overhead. Cirrus clouds are usually more than 7,000 metres (say about 20,000 ft) above the surface of the earth.

Sometimes the upper air is completely filled with an ice particle cloud which is so thin as to be practically invisible. The presence of such a cloud may only be made known to us by the halo which forms in it, around the sun or the moon. Such a cloud producing a very clear halo is shown in Colour Plate 2. Careful observation shows that under such conditions the sky appears paler than the normal clear blue of fine weather. The sky is darker inside the halo than outside. Although the name cirrus is quite unsuitable for such a cloud, since there are no fibres to be seen in it at all, it is still used in that case, because the cloud is high and made of ice particles. Sometimes it is called a cirrus nebulosus cloud or a cirro-nebula, the prefix cirro being the adjective derived from cirrus, and nebula meaning a misty cloud.

Usually lower in the sky, a very common form of cloud, known as cumulus, develops. The word cumulus means a heap, and cumulus clouds take the shape of heaps of clouds, apparently piled on top of each other. Colour Plate 5 shows an example of a well-developed heap. Like cirrus, cumulus clouds can form at a variety of heights. When very high indeed, and made of ice particles, they are called cirro-cumulus clouds. Below the level at which cirro-cumulus clouds usually form, but still very high in the sky – from about 2,300 to 6,500 metres (7,000 to 20,000 ft) – a very beautiful type of cloud is often to be seen. It is known as alto-cumulus, the prefix meaning high. Alto-cumulus consists of apparently small cumulus clouds, often arranged in patterns, and forms the mackerel sky of the proverb 'Mackerel sky, rain nearby' – as does cirro-cumulus also. Colour Plate 3 is a photograph of such clouds at sunset. Being high in the atmosphere they catch the last rays of the setting sun, and the first rays at sunrise. Colour Plate 4 is a photograph of alto-cumulus cloud as seen from above. The photograph was taken from a height of 11,000 metres (33,000 ft) in the centre of the USA, on a flight from Washington to San Francisco.

Any more or less uniform, horizontal layer of cloud is known as stratus. The high cloud in Colour Plate 2 giving rise to the halo, which we have called cirro-nebula, could have been called cirro-stratus, though the term is usually reserved for layers of cirrus clouds which are more obviously to be seen. Colour Plate 6 shows a layer of stratus, which is much lower though still high. There is no halo of 22° although the sun looks very 'watery'. There are no ice particles at this level. The sun, however, is surrounded by a 'diffraction' halo, which will be discussed in Chapter XI, and this gives it the watery appearance.

Colour Plate 7 shows stratus at a lower level still. A fog is also a form of stratus cloud – it is nothing more than a cloud at ground level. Sometimes the depth of the fog is quite small and the tops of the trees and high buildings are clear, as in Colour Plate 8. By going out into a fog we can appreciate what the inside of a cloud is like. It is not very pleasant. The air is damp and filled with a host of tiny water droplets. Some of these get caught on the hairs of the cloth of our overcoats. Some are too small to be visible as they are, while others join up to make droplets which are big enough to see.

There is one other term in common use to describe clouds and that is nimbus. This means any cloud from which rain or other type of precipitation is falling. At first it is necessary to take care not to confuse the descent of rain with rays of sunlight penetrating gaps in the cloud and showing up as light and dark streamers beneath the clouds. It is not, however, usually difficult to distinguish between the two. Rays of sunlight radiate from the sun and are straight, whereas falling patches of rain are usually not so. A surer way of telling that it is rain is to notice whether or not it masks the distant landscape. Colour Plate 9 shows rain obscuring the distant mountains. The view is of the Cuillin Mountains in the Isle of Skye, photographed from near Elgol. Colour Plate 20 is the same view taken from near the same spot on a different occasion when the more distant mountains were visible. It is one of the finest views of mountain scenery in the British Isles. A coloured rainbow – Colour Plate 10 – is a sure indication of the presence of raindrops.

Besides the terms so far used, a number of others can be made by joining two of the first kind together. The first term is normally used in its adjectival form ending with the letter 'o'. We have already met with cirro-nebula. Cirro-cumulus, for example, would be a cirrus cloud possessing the heaped form of the cumulus cloud. In the case of cirro-cumulus, the heaps appear very small, the cloud having a flocculent appearance. A very common combination is cumulo-nimbus – in its shortened form Cu–Nb – which is a cumulus cloud from which rain or snow is falling. Thunderstorms are the larger forms of cumulo-nimbus. Stratus cloud from which rain or snow is falling is known as nimbo-stratus. Sometimes the adjectival part of the name is placed second, as in stratus cumuliformis. This is stratus cloud having a lumpy character, like cumulus. There should be little difficulty over these combinations of terms. Their meaning is usually quite clear.

There are recognised abbreviations for cloud names which are worth remembering for the writing of notes. The abbreviation for cirrus is Ci., that for cumulus, Cu., that for stratus, St. and that for nimbus, Nb. Alto-cumulus is shortened to A.Cu. Other combinations can be written down easily. Thus strato-cumulus is shortened to St.Cu.

After we have studied some of the processes that go on in the atmosphere, we shall find a number of other features to look for in observing clouds. These will require a few more names to be learned. It will be better, however, to leave them until we know a little more of what goes on in the air. They will become clear enough as each case is considered.

In looking at cumulus clouds, one of the most important points to look for is the amount of what is known as vertical development which is present, especially if one is interested in the question of whether or not it is likely to rain. Vertical development indicates instability in the atmosphere, as we shall see in greater detail later, and leads to the formation of rain clouds. Colour Plates 11 and 12 show two contrasting cases. Colour Plate 11 is a photograph of some cumulus clouds taken at 10.30 in the morning GMT, showing marked vertical development. Heavy showers were falling a little over an hour later. In Colour Plate 12 on the other hand, we see cumulus clouds taken at half past one in the afternoon on a summer's day. The weather was very settled and the clouds show very little vertical development. Early vertical development is a sign of the likelihood of heavy showers a little later in the day.

We will now turn our attention to a little study of the physics of the atmosphere. To do this completely would mean studying nearly the whole of physics. We shall have to remain content to examine only a few of the most important items.

Chapter II

Water Vapour in the Atmosphere

It is not always understood what the term 'vapour' means. Nearly all substances can exist in three states – the solid, the liquid, and the gaseous. These states are well known and there is no difficulty in distinguishing them. Solids maintain a nearly constant shape and size. They may be bent, or expanded a little by heat, but otherwise their shape and size remain unchanging. Liquids, on the other hand, possess a definite size or volume, but no definite shape. They can take up any shape – usually that of the vessel which contains them. Gases differ still further in that they have neither definite size nor shape. A quantity of gas can fill any vessel in which it is placed, however large or however small it may be.

When a liquid evaporates its substance changes from the liquid to the gaseous form, and when it condenses again it goes back to the liquid form. All gases can be liquefied (or condensed). Those which can be condensed relatively easily by simply increasing the pressure on them, are known as vapours. The difference between a gas and a vapour is thus not very great. The term gas is sometimes used to mean only those which can be condensed with difficulty (the so-called permanent gases) but it is better to use the word gas to include all bodies in the gaseous state. Gases thus include vapours, and vapours are only certain kinds of gas.

The majority of the common gases, such as the oxygen and nitrogen of the air, and the carbon dioxide which is also present in small quantities, are colourless and invisible. Certain gases, like chlorine and nitrogen peroxide, are coloured and can be seen easily. Water vapour, however, is one of the colourless and invisible gases. Steam, properly so-called, is water vapour, and it too is invisible. What we see coming out of the spout of a kettle of boiling water, cannot, therefore, be steam. It is simply a cloud of tiny droplets of ordinary water, which have condensed from the hot steam or water vapour coming out of the spout, as it mixed with the cool air outside. The 'steam' coming in dense clouds from the old steam railway engines (Colour Plate 13) is similarly not steam. Some of it is smoke from the burning coal, but much of it is, again, merely a cloud of water droplets.

 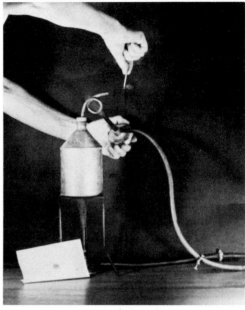

Plate II.1 (A)   Water is being boiled in the can and the familiar cloud, usually misnamed 'steam', comes out of the copper tube. The cloud is not really steam but a host of tiny droplets of liquid water, which have condensed when the steam, which is invisible, mixed with the cold air outside. Near the end of the tube there is a clear space before this mixing and cooling has taken place.

Plate II.1 (B)   If now the copper tube is heated with a second bunsen burner, the steam can be made so hot that it is not cooled sufficiently to condense before it has mixed with a very large amount of air, and condensation does not occur. The cloud then disappears. The presence of hot water vapour (superheated steam), however, can be demonstrated by holding a piece of paper in its path. The paper soon shows a dark scorch mark. More dramatically, a match held in the hot vapour will ignite.

A simple experiment will serve to demonstrate these facts. If we boil water in a can and allow the 'steam' to come out through a spiral of copper tube, as in Plate II 1 (A), we repeat the familiar experience of the kettle of boiling water. We see the usual cloud, commonly misnamed steam, coming out of the tube. Near the end of the tube, however, where the water vapour is still hot, there is a clear space. In this space we have true water vapour or steam. The clear space can also be seen with an ordinary kettle if the water is boiling vigorously. The water vapour has not cooled sufficiently to condense into the cloud of water droplets and it is invisible. If we heat the copper tube with a second bunsen burner, the water vapour coming out of the can can be made so hot that it does not condense until it has mixed with so much air that the cloud is never formed at all. The air is able to hold a good deal of water vapour as a gas, as we shall shortly see. However, the presence of the stream of very hot water vapour (superheated steam as it is sometimes called) coming out of the tube can easily be demonstrated. A piece of paper held near the end of the tube is soon scorched by it and it is even hot enough to ignite a match, as the photograph in Plate II 1 (B) illustrates.

It is very easy to show that under ordinary conditions there is already water vapour in the air. It exists as one of the gases of the air. If air is cooled sufficiently its water vapour will condense into the liquid form. It is a familiar experience that a jug of cold water brought into a warm room soon gets covered with drops of water on the outside. These drops have come from the water vapour in the room, and are really like drops of dew, which are formed in the same way by cooling at night out of doors. Water vapour will also condense on any body brought indoors on a cold winter's day, and if the body is cold enough the condensed water will freeze to ice. When the body is very cold, however, the water vapour will condense on to it directly in the form of ice, without passing through the liquid state on the way. This is what hoar frost is. It is composed of many small crystals of ice and, like sugar or salt which are also composed of many small crystals, it looks white. Black ice, on the other hand, is formed by the freezing of water which has first condensed as dew – or has fallen on the ground as rain.

The can in Plate II 2 contains a freezing mixture of ice and salt, and its temperature is much below freezing-point. The white deposit on the lower part, where the freezing mixture has cooled the metal of the tin, is hoar frost, condensed directly as the solid out of the atmosphere. The can stands on a block of wood which has been smartened up to be photographed by being covered with white paper. It is not ice though it looks the same colour as the hoar frost!

Plate II.2   A freezing mixture of ice and salt has been placed in the can. Its temperature is well below freezing-point. The water vapour in the outside air has condensed on to the lower part of the can, where the metal has been cooled by the freezing mixture. It condenses directly in the form of ice without forming liquid water first. The deposit appears white because it is composed of a multitude of tiny ice crystals. Hoar frost is formed out of doors in winter in the same way.

The air can only hold a certain amount of water vapour and when it can take no more it is said to be 'saturated'. In saturated air no further evaporation of water can, of course, occur. Clothes will not dry and the body cannot cool itself by the evaporation of perspiration. In summer the weather feels unpleasant, close, and oppressive. In winter it feels cold and 'clammy'.

Certain substances absorb water vapour from the air. Ordinary table salt does so to some extent and goes damp if left exposed for long. Strong sulphuric acid, calcium chloride, and phosphorus pentoxide do so much more. By weighing a little of one of these substances in dry air and then passing a known volume of saturated air slowly over it, and weighing it again at the end, it is possible to find out how much water vapour a given quantity of air can hold. It is found that the amount increases very much as the air gets warmer. Very cold air is necessarily very dry and in countries which have very cold winters, such as North America, the outside air contains so little moisture that when it enters buildings it is so dry that objects readily get charged electrically. You yourself may become charged, for example, by walking across a carpet and draw sparks off the door knob when you open a door.

Some objects change in size with the amount of water vapour in the air. Tracing paper fastened across a frame to make a translucent screen for the lantern, for example, may be tight as a drum when the air is dry, but loose and baggy when it is damp. Catgut, fastened at one end and stretched by a weight, twists and untwists as the dampness of the air changes. This effect is used to make the old-fashioned 'weather châlet', the man coming out when the air is damp and the woman when it is dry. The air is not always damp, however, when it is about to rain, nor is it dry always in fine weather, so that the châlet is of limited use as a forecaster of the weather. A hair alters its length and one end can be fastened and the other attached to a lever to move a hand over a scale. The scale can be graduated to read the 'humidity'. This is the amount of water vapour in the air at the time, expressed as a fraction of the amount required to saturate it at the same temperature. A humidity of 60 per cent, for example, means that the air is carrying 60 per cent of the maximum amount of water vapour which it is capable of carrying at the same temperature.

Whenever the temperature of the air falls sufficiently the water vapour it contains may condense to form dew, hoar frost, or a cloud of water droplets. Our study of clouds must, therefore, next take us to consider the ways in which air can be cooled. As we might expect there are, in fact, several ways in which the air can be cooled sufficiently for its water vapour to condense, but there is one way which is of outstanding importance. It is particularly effective because it can operate inside a large volume of air, and in this it differs from contact with a cold surface, for

example, which only cools the layers at the bottom quickly, the rest of
the air taking a much longer time to cool unless the air is stirred up by
some means.

When a bicycle tyre is being pumped up the end of the pump, where
the air is compressed, gets warm. It seems at least possible that when air
is allowed to expand, the opposite happens, so that it cools. It is easy to
show that this is so. Fig. II.1 shows a diagram of a simple apparatus for
the purpose. The vessel is an oil drum and it is closed and made airtight
with a rubber bung through which two glass tubes pass. One is connected
inside the drum, by means of another rubber bung, to a 7·5 by 2·5 cm
specimen tube. The object is to make an air thermometer to indicate the
temperature of the air in the drum. The outside end of the thermometer
tube is connected, by means of a piece of rubber tubing, to a horizontal
glass capillary tube, which forms the stem of the thermometer. Its end
is bent downward and dips into a tube of coloured water. By squeezing
the rubber tubing, a little of the air in the thermometer can be driven out
and the coloured water made to enter the end of the thermometer stem.

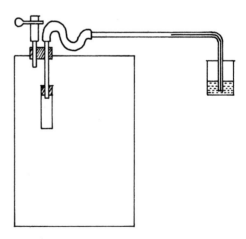

Fig. II.1

The coloured water is merely used to indicate the end of the column of air in the stem of the thermometer. If the temperature of the air in the drum falls, the coloured water will move to the left and if it rises it will move to the right. The five-gallon oil drum and the scale of the air thermometer can be seen in the photograph in Plate II.3 (A). The other tube in the bung closing the oil drum is used to compress the air inside a little. Oil drums will take quite safely about fifty strokes with a bicycle pump fitted with a football adapter, but it is best to wire the bung to the drum to hold it in place to prevent it from blowing out. Fifty strokes with an ordinary bicycle pump will deliver about two and a half litres and the pressure in the drum will rise by about an eighth of an atmosphere.

Plate II.3 (A)   The photograph is of a five-gallon oil drum to which a simple air thermometer has been fitted. The end of the column of air in the horizontal stem of the thermometer is marked by coloured water, which has been drawn into the tube by pinching and then releasing the rubber connecting tube.

Plate II.3 (B) As a precaution against accidents the drum has been placed in the tea chest and fastened down, the connecting tube of the thermometer being taken through a hole in the lid. The drum is pumped up with about fifty strokes of a bicycle pump fitted with a football adapter. The clip is closed and the apparatus allowed to stand for a few minutes. The photograph shows what happens when the clip is then released. The air expands and its temperature, as indicated by the air thermometer, is seen to fall.

Though this should be quite safe, as a precaution the drum is placed in a tea chest and the tubes are taken through a hole in the lid, which is fastened down, before pumping up. The drum is then pumped up, the clip closed and the thermometer set to register near the end of the capillary tube. After a short interval to allow the temperature to become steady, the clip is opened again to allow the extra air which has been introduced into the drum to escape. When this happens the air in the drum expands and the air thermometer shows that the temperature falls, as indicated on the photograph in Plate II.3 (B).

A word of caution is necessary about compressing air in any closed vessel. Oil drums are strong and can be pumped up safely if high pressures are not desired. Using a five-gallon drum, more than about a hundred strokes with an ordinary bicycle pump fitted with a football adapter should not be used without a safety valve, even with the drum in a box. A simple form of this is a U-tube half filled with mercury, as in Fig. II.2. The end of the U-tube is bent over so that any mercury expelled can be caught in the flask. It should be set to blow at about a quarter of an atmosphere pressure, which will be the case if the mercury surface to start with is about 10 cm below the top of the bend. It is a wise precaution, in any case, to put any drum in which the pressure is to be increased in a tea chest fitted with a lid through which the necessary tubes can be led.

Glass vessels should not be subjected to more than very modest pressure increases and air in them should only be compressed by direct blowing with the mouth.

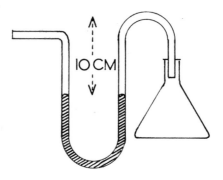

Fig. II.2

A rather more spectacular way of showing the cooling of gases on expansion is to use the little cylinders of highly compressed carbon dioxide which are sold to make siphons of soda water. If the seal is punctured with a small hole, the gas expands rapidly and the cooling can be very considerable. Plate II.4 (A) shows the cylinder, the pointed instrument, such as a piece of sharpened knitting needle, and the hammer with which to hit it. The black cloth is used to surround the opening before hitting the needle so that as the gas comes out it passes between the folds in the cloth. The reason for doing this is that the cooling may well be sufficient to solidify the carbon dioxide gas and Plate II 4 (B) shows the result. The white deposit on the cloth is solid carbon dioxide. It quickly evaporates. The cylinder becomes so cold that hoar frost is deposited on it, and it has frozen a small drop of water on the matchbox so that it becomes stuck to it.

Cooling by expansion is a most important process in the atmosphere itself, and these two simple experiments suggest a method by which we might try to make a cloud artificially in the laboratory. Suppose that we take a glass jar full of damp air. The jar in Plate II.5 (A) has a little water in the bottom to keep the air moist. The pressure inside the jar is raised slightly by blowing into the rubber connecting tube as hard as we can by mouth. The tube is then nipped in the fingers and held for a moment or two to allow the temperature to become reasonably steady. This is the

Plate II.4 (A)   Small cylinders of highly compressed carbon dioxide are sold for the making of siphons of ' soda water' at home. They are sealed by a blob of solder, which can be pierced by a sharp stout needle hit by a hammer. When the seal is pierced there is a rapid expansion of the carbon dioxide. The black cloth is to surround the opening before striking the needle. The matchbox has a small drop of water on it.

stage taken in the photograph of Plate II.5 (A). On releasing the pressure a cloud is distinctly seen to form inside the jar. Plate II.5 (B) corresponds to this operation. In Plate II.5 (C) the experiment has been repeated, but before doing so a little smoke was introduced into the jar by burning a piece of brown paper near the mouth of the jar. When the air is again expanded a very much denser cloud is produced, as Plate II.5 (C) shows. Clouds and fogs readily condense on smoke particles, and to prevent the occurrence of bad fogs in cities many local authorities have instituted smokeless zones, in which only fuels which produce no smoke are allowed to be burned. Before this had been done most large cities, such as London and Manchester, had their own 'particulars'. A bad one which happened in Manchester, in which it was impossible to see the ground on which one was walking, makes a vivid memory. Visibility ended at about knee level on that occasion!

Plate II.4 (B)   The seal has been pierced and the carbon dioxide has been cooled by expansion to such an extent that some has been deposited as a solid on the black cloth. It quickly evaporates. Hoar frost is deposited on the cylinder, which has frozen the drop of water on the matchbox so that it sticks to it.

There remains one question which might be asked in this chapter. It is 'If clouds are made of drops of ordinary water, and not water vapour, why do they not fall down? After all rain, which is also made up of drops of liquid water, falls down quite quickly.' The answer is that clouds do, in fact, fall through the air all the time, but that their rate of fall, even if the air were still, which it rarely is, is so slow as to be usually imperceptible. This is because the droplets of which clouds are composed are very small. The air slows up the rate of fall of small bodies, as can easily be seen by dropping some very fine sawdust at the same time as a large piece of the wood from which it came. The piece of wood reaches the ground long before the sawdust. In the early days of physics an experiment used to be performed to show the effect of the air on falling bodies. It was known as the 'feather and guinea experiment'. A feather and a guinea piece were allowed to fall inside a long glass tube. The guinea piece reached the end of the tube while the feather had drifted down only a small part of the way. The air in the tube was then pumped out and the experiment repeated. It was found that when this was done both the feather and the guinea piece fell together at the same rate. It was the action of the air on the feather which made it fall so slowly.

If the air happens to be moving upwards very fine particles may be carried upwards instead of falling. If we have been moving about in the room before performing the experiment with the sawdust, and if the sawdust is very fine, we may well see some of the dust carried upwards. Mahogany produces very fine sawdust and is suitable for the experiment. Thistledown, though it is, of course, heavier than air and will finally settle somewhere, can be carried to considerable heights by currents in the air, and transported long distances before finally coming down to earth. On 1st July 1968 fine dust from the Sahara was carried over England and was deposited on houses and other objects there.

The droplets of water in a typical cloud range in size from two-hundredths of a millimetre in diameter to about twenty times this size. The rate of fall of the smallest drops amounts to only about a millimetre a second. It would take a quarter of an hour for them to fall a metre, so that the time for them to fall even 300 metres (about 1,000 ft, say) would be over three days, long before which the cloud itself may well have dispersed altogether. Drops 2 mm in diameter, on the other hand, which are raindrops rather than cloud droplets, fall at a speed of about 650 cm a second in still air, once they have reached their final velocity.

Plate II.5 (A)   The jar contains air which is kept moist by a little water in the bottom. The air inside has been compressed by blowing into the connecting tube as hard as possible by mouth. The tube is then nipped in the fingers, and held for a moment or two for the temperature to become steady.

Plate II.5 (B)   On releasing the pressure a light mist is seen to form inside the jar. The cooling produced by the expansion of the air has been sufficient for a little of the water vapour in the air to condense into drops of liquid water.

Plate II.5 (C)   The experiment has been repeated but before doing so a little smoke was introduced into the jar from a piece of burning brown paper. The expansion of roughly the same amount of air as before produces a much denser cloud.

Chapter III

Air has Weight

The subject of this chapter takes us back to the very early days of physics, just after the nature of gases had come to be appreciated. It was not realised then that gases, like all other bodies on the earth, possess weight. The period when this was discovered to be the case dates from around the time of the foundation of the Royal Society in 1660. Indeed the early members of the Society were chivied for having 'done nothing but weigh air since they sat'. The result was hard to come by because the necessary techniques of weighing and the manufacture of efficient air pumps were only then being developed. Galileo had obtained a value somewhat earlier than this, using the method of which we shall also avail ourselves. His result, which he described in his book called the *Dialogue Concerning Two New Sciences*, which he completed in 1636 and published in 1638, was not very accurate, being little more than half the accepted value. So long as we remain content with rough results the fundamental facts are easy to demonstrate using a very simple, home-made apparatus.

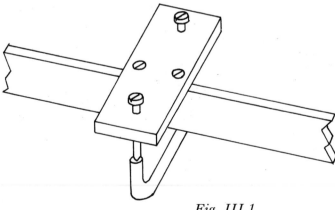

Fig. III.1

Plate III.1   A colourless heavy vapour (carbon tetrachloride) being poured into a plastic bag hung on a simple balance. The beam is first balanced and the can filled with vapour by warming a cc or two of the liquid in it (the top being closed with a card). When the vapour is poured into the bag nothing will be seen to emerge but the balance will go down in no uncertain manner. The heavy vapour can be poured out of the bag by tilting it, just like a liquid; the balance will then be restored and the experiment can be repeated.

To do so we can make use of a simple demonstration balance. A photograph of one is shown in Plate III.1. It consists simply of a lath of wood about a metre long balanced on two screws sharpened to points, which are attached to it by a cross piece. The pointed ends of the screws sit in depressions in the metal rod which supports the beam. Fig. III.1 is a diagram of the arrangement. The advantage of having the supporting points on the ends of the screws is that their height can be adjusted. The sensitivity of the balance depends upon how high the beam is mounted relative to the supports, and by adjusting the screws the sensitivity of the balance can be varied. The beam is balanced by means of a U-shaped piece of wire which, if inverted, can be hung on the beam and moved in position until a balance is obtained. The sensitivity should be adjusted by setting the screws at such a height that the balance is completely deflected by a small weight of about a quarter of a gramme.

An amusing preliminary demonstration which can be carried out with the balance is the pouring of invisible heavy vapours from one vessel to another. This is shown being performed in Plate III.1. Plastic bags have been hung on the balance for the purpose, but paper ones, kept open at the top with a fine wire ring, would serve equally well. The beam is then balanced by adjusting the position of the rider. The metal can is filled with some heavy vapour. We need for the experiment a heavy, invisible and preferably non-inflammable vapour. Trichlorethylene or carbon tetrachloride vapours serve very well. Both of these compounds are sold in chemists' shops as liquids for cleaning clothes. Carbon tetrachloride is rather more poisonous than trichlorethylene, but there is no need to breathe either of the vapours in doing the experiment. A few drops of the liquid are placed in a large can. The top is covered with a piece of cardboard and the vessel is warmed gently over the gas. It soon fills with the vapour of the liquid. The vapour can then be poured into the bag on the balance. Nothing will be seen to come out of the can, but the balance will go down with quite a bump, and remain down for a long time. The vapour can be emptied out of the balance bag, as though it were a liquid, by turning the bag upside down. The balance will then be restored to equilibrium if care has been taken not to disturb the position of the rider. The experiment can then be repeated. The heavy vapour can first be poured from one vessel to another as a preliminary before emptying the second into the bag on the balance, if desired. The whole experiment seems a little ludicrous and amusing, pouring apparently nothing from one vessel to another.

We have said that nothing can be seen coming out of the can of vapour when it is poured. This is not strictly the case. If one looks carefully it is possible to detect the stream of vapour as it is poured out of the can. It looks like the currents of hot air which can be seen rising from a heated wall or pavement on a hot day. A similar stream of heavy vapour can be seen as the petrol tank of a motor car is being filled. The refractive index of the heavy vapour or the hot air is slightly different from that of the surrounding air, and this makes the background appear distorted. A similar distortion of the background is produced by the hot gases which

come from the exhaust of the jet engines of an aeroplane, and the view
of the ground is spoilt in this position. A better view is obtained in front
of, rather than behind, the engines.

Extreme caution is desirable when dealing with heavy vapours which
are inflammable, such as the vapours of petrol or ether. The streams of
vapour descend rapidly to the ground. Once on the ground they flow
along it just as a liquid would do, and it is necessary to be extremely
wary and avoid all flames in the neighbourhood. A cigarette end or a
lighted match, thrown down at a considerable distance from a car when
the petrol tank is being filled, can cause the vapour to ignite. The flame
runs along the stream of vapour on the ground to the car and disaster
can result. A friend of the author had this actual experience with one of
the old type Austin motor cars in which the petrol tank was underneath
the front seat. (In the evolution of the motor car the petrol tank started
behind the dash-board where the driver and passenger practically
nursed it on their laps. It then moved amidships, as in the car described,
before finally reaching its present position at the rear. The moves were
determined by questions of safety.) Fortunately he was able to jump
clear, but a wall of railway sleepers near the car was burnt out completely
by the resulting fire, which, of course, also destroyed the car and its
contents. Similar flames can run along the floor when inflammable
vapours are used in the home. They are given off by certain adhesives,
for example, and by benzene used for cleaning clothes, but there is now
no need to employ inflammable liquids for the latter purpose. Similarly,
great care is necessary in the laboratory when using ether, which also
gives off a heavy inflammable vapour, and a flame can run across a
laboratory floor or bench. Town gas, which is lighter than air, on the
other hand, will accumulate under the ceiling, and can also ignite there
and carry a flame from one part of a building to another. Fire damp in
mines also accumulates under the top surfaces of passages and, if it
becomes ignited at one point, can cause explosions in another part of the
workings.

The simple demonstration balance can be used to show the weight of
air itself. The experiment is shown in Plate III.2. Extra air is pumped
into the bottle by means of a bicycle pump fitted with a football adapter
and the beam is balanced by means of the rider. In this case we need to
raise the pressure inside the bottle by an amount which would not be at
all safe with a glass one. Accordingly, for this experiment we choose a

plastic bottle, such as certain kinds of lemonade are sold in. The rubber
bung is wired firmly into place and we need to pump about half a litre
of extra air into the bottle. This requires about ten strokes with the cycle
pump. Depending on the size of the bottle this may raise the pressure
inside by as much as one atmosphere. When the extra air has been pumped
in, the clip is closed on the connecting tube and the beam is balanced. If
a quantitative result is not required there is no need to do this carefully,
since we shall have over half a gramme for the balance to indicate. Plate
III.2 (A) shows the bottle with the extra air in it balancing well down on
its own side. When the clip is carefully opened so as not to disturb the
rider, and the extra air is let out, the balance swings over, as shown in
Plate III.2 (B), because of the weight of the air which has left the bottle.

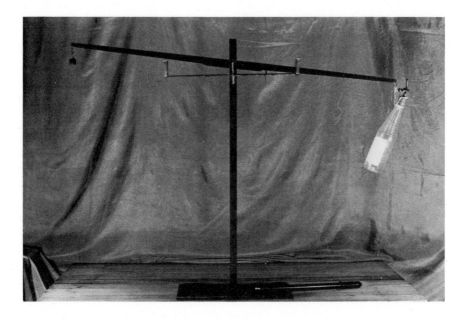

*Plate III.2 (A)   A <u>plastic</u> bottle (a glass one must on no account be used
for this experiment) has a rubber bung wired into its mouth, and extra air
is pumped in – giving ten strokes with the bicycle pump fitted with the
football adapter. The beam is first roughly balanced, being well down on
the bottle side.*

Plate III.2 (B)   The extra air is let out by carefully releasing the clip so that the position of the rider on the balance beam is not disturbed. The loss of the weight of the extra air is clearly indicated by the beam, which swings well over to the other side.

By placing weights on the beam of the balance just above the bottle, the weight of the lost air can be found. If the bottle is filled again by means of the cycle pump, using the same number of strokes as before, the volume of the extra air which has been introduced can be measured by connecting the rubber tube to a glass delivery tube, so that when the clip is opened the extra air will collect in a measuring cylinder inverted over water. In this way the weight of a litre of air can be measured. Although the experiment is only rough considered as a quantitative measurement, values not far removed from the accepted value for the weight of air will be obtained all the same. Dry air weighs from about 1·2 to 1·3 g/l, according to the temperature and pressure. The weight is less the higher the temperature and the lower the pressure. Damp air weighs slightly less than dry air.

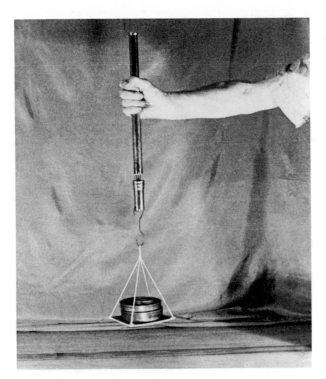

Plate III.3   Measurement of the atmospheric pressure by means of a bicycle pump. The plunger inside has been reversed so that the pump sucks instead of blowing, and a good seal is achieved by lubricating it with a little engine oil. The mouth of the pump has been closed. When the handle is pulled out the pressure of the atmosphere causes it to fly back again as soon as it is released. It requires considerable weights hung on the handle to pull it out. Here the piston is seen supporting 3 kg (6½ lb). The area of cross-section of the pump is 3·22 cm² (½ in²) so that this measures the atmospheric pressure to be about 6 kg (13 lb) per 6·45 cm² (in²), or just under 1 kg per cm². Actually it will be somewhat greater since there is the weight of the handle and of the scale pan to be added on.

This weight of air is quite considerable. An ordinary classroom, 4 m (about 12 ft) high and 45 m² (about 500 ft²) in area contains about 230 kg (nearly 4 cwt) of air. This is nearly one third of the weight of a class of 30 pupils each weighing 25 kg (about 4 st.). But we live at the bottom of an ocean of air, the weight of which has to be carried by the surface of the earth. It is thus only to be expected that the air will exert a very considerable pressure. It is well known that the pressure of the atmosphere is about 1·033 kg/cm² (which is 14·7 lb/in²). This is a very large force for such a small area, and a simple experiment or two to bring the size of this force before the imagination, can be well worth doing.

Laboratory suppliers sell specially thin metal cans which can be exhausted by means of an air pump. When the air is removed from inside, the pressure on the outside causes the can to collapse. Since the can is especially thin the result is not of great help. However, the same experiment can be done in a much more spectacular manner using an oil drum. These are strong vessels on which it is possible to stand, so that any force which is capable of making it collapse must be of considerable size. The air can be pumped out of it by means of an air pump and the drum will not prove sufficiently strong to withstand the external pressure, which is no longer balanced by the pressure of the air inside. Alternatively, some water can be boiled in the bottom and the drum filled with steam. The flame is taken away and the drum closed with a rubber bung. When the drum cools the steam condenses and there is a vacuum left inside, and a similar collapse ensues.

The value of the pressure of the atmosphere in kilogrammes weight per square centimetre (or pounds weight per square inch) can be obtained roughly by means of very simple apparatus. The plunger of a bicycle pump is usually made of leather in the form of a cup. The cup faces forward so that when it is pushed so as to pump up the tyre, the sides are pressed against the walls of the pump and make an airtight joint. If this leather plunger is reversed, so that the cup points backwards (Fig. III.2), the pump will suck instead of blow. To measure the pressure of the atmosphere, a little engine oil is placed on the plunger to make a good seal and the opening of the pump to which the connection is usually attached is closed up (it can simply be closed by the thumb if desired). If the handle of the pump is now pulled out and then released, it flies back in again because of the force exerted by the atmosphere on the plunger. The vigour with which it does so is itself a striking illustration of the size of the pressure of the atmosphere. However, to obtain a quantitative value for the atmospheric pressure the pump is held vertical with the handle pointing downwards. A scale pan is attached to the handle and on it are placed weights. At first the handle remains in position, but as the weights are increased there comes a point when the pressure of the atmosphere is no longer able to hold them up and the handle descends. By adjustment, the weight which is just sufficient to overcome the pressure of the atmosphere can be determined. Knowing

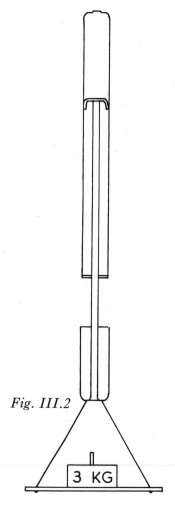

Fig. III.2

3 KG

this, the pressure of the atmosphere can be calculated. Plate III.3 shows a pump supporting about 3 kg (actually $6\frac{1}{2}$ lb) in this way. The area of cross-section of this pump was 3·2 cm² ($\frac{1}{2}$ in²) so that the air must exert a force on this area sufficient to support the 3 kg, together with the weight of the handle, hook and scale pan. The handle and its attachments weigh about 100 g, so that the experiment indicates that the atmospheric pressure must be about 1 kg weight per square centimetre (14 lb/in²). The value is a little below the one which is usually accepted but the experiment is not capable of being made very accurate.

A pressure of this size exerts a very considerable force on any extended area. It was this force which caused the oil drum to collapse when the air inside it, which presses outwards, was removed. On each square metre of surface there is a force of about 10 tonnes. On the whole surface of the earth the force which is exerted must be equal to the weight of the atmosphere. This amounts to some $5\frac{1}{4}$ thousand million million tonnes.

Vessels placed at a depth in any fluid – liquid or gaseous – experience similar pressures. The effect in water is much greater than in air because water is so much more dense than air. Nine metres (30 ft) of water produce the same pressure as the entire atmosphere. At the bottom of the oceanic depths the pressure can amount to a thousand atmospheres – 10,000 tonnes on every square metre. Creatures can become acclimatised to the pressures at these great depths because their bodies are diffused with water and gases at the same pressure, just as an open can will exist without collapsing. If, however, a vessel with air at atmospheric pressure only inside it were lowered to these depths, it would collapse under the pressure unless it were made enormously strong. Ordinary submarines are strictly limited to the depths to which they can dive without risking collapse from the external pressure of the water. Those specially designed to go down to great depths have to possess extraordinarily strong hulls. Conversely, if creatures are dredged from the sea bed at considerable depth, gases dissolved in their body fluids will be released and will expand and the creature may blow up. Divers experience a similar occurrence when brought to the surface after spending some time at considerable depth. The air in their diving suits has to be forced to them under a pressure corresponding to the depth at which they are working and some of it dissolves in their blood. When the pressure is released the gas comes out again in the same way that it comes out of 'soda water' when the pressure is released, and it can cause blockages in the circulatory system. The condition is known as 'the bends', and to avoid it divers must be brought to the surface slowly or put into air at high pressure in a decompression chamber, in which the pressure can be lowered gradually. The pressure in the middle of stars can reach unimaginable proportions and matter can become exceedingly condensed, as in stars of the type known as white dwarfs. The density of the star Sirius B – the companion of Sirius – has been estimated to be 50,000 times that of water.

Plate III.4   The Puy de Dôme. This is the mountain to the top of which Pascal's brother-in-law first carried a barometer to measure the atmospheric pressure. Today the summit is disfigured by a gigantic tower which almost dwarfs the mountain. In this photograph the tower is fortunately largely obscured by clouds.

Consideration of the next fact with which we must be familiar in thinking about the formation of the clouds in the atmosphere, again takes us back to the seventeenth century – to the year 1646 in fact. In that year Blaise Pascal, who lived at that time in Paris, asked his brother-in-law, a Monsieur Périer of Clermont, to take a mercury barometer to the top of the Puy de Dôme, a mountain 1465 metres high, a short distance to the west of the city. This he did most carefully, first setting up a barometer in the lowest part he could find nearby, and getting an observer to watch it all day for changes, while he himself took another to the top of the mountain. He measured the height of the barometer at several points near the summit and on the way down, and was able to show conclusively that the pressure of the atmosphere was less at the top than at the bottom. At points between the top and the bottom it had an intermediate value. The experiment helped people to understand that the pressure of the atmosphere was caused by the weight of the air above them. The next day a barometer was taken to the top of the highest tower of the Church of Notre Dame de Clermont, and a small but measurable decrease in the height of the mercury was observed. Even at the beginning of the nineteenth century Faraday, when on his travels with Sir Humphrey Davy, thought it worth while to observe this fact of nature for himself, and when they crossed from France into Italy in mid-winter in 1814, he himself carried on foot a mercury barometer and a thermometer. He recorded the temperature on the top of the Col de Tende to be $-12°$ C ($12°$ C or $21°$ F of frost) and the height of the barometer to be 64 cm of mercury. We too can see the phenomenon for ourselves and can do it much more easily and conveniently than Faraday, since we can use an aneroid barometer in place of one of mercury. The barometer falls about 8 mm for every hundred metres of ascent (about an inch per thousand feet), so that if we can get up a tower or hill even 30 m high (say about 100 ft) we should see a difference of almost 3 mm (about $\frac{1}{10}$ in). We can even do the experiment without going out of doors, in one room, using a height of only a metre or so.

Connect a wide tube (about two or three centimetres in diameter – say about an inch) to an empty oil drum (preferably a ten-gallon drum – the larger the better) by means of a piece of rubber tubing, and put a soap film across the end (Fig. III.3). It will lie flat across the mouth of the tube. Then lift up the oil drum and tube together about a metre and the soap film will blow out because the pressure of the air in the drum will be greater than the pressure of the atmosphere in the higher position. The difference is sufficient to make the soap film bulge out unmistakably.

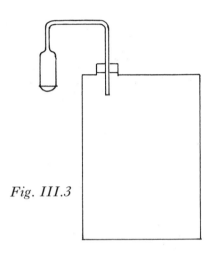

Plate III.5 (A)   An experiment to show that the pressure of the atmosphere decreases with height. Inside the tea chest is a ten-gallon oil drum, packed round with paper and straw to diminish the effects of changes of temperature. The drum is connected to a glass tube across the mouth of which a soap film has been placed. In Plate (A) the drum is under the table and the soap film is level with the end of the tube.

Fig. III.3

(A)

In Plate (B) the drum
has been lifted on to the
table and, because the
presence of air inside the
drum is now greater than
that of the surrounding
air, the bubble blows out.

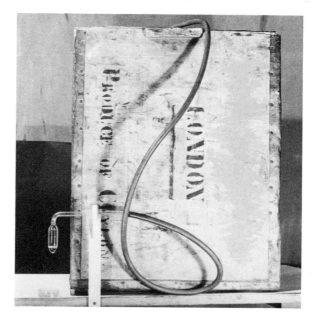

(B)

The actual experiment is illustrated in Plate III.5 (B). In this demonstration the soap film is left at the same level throughout, but it makes no
difference to the result. It is an interesting little problem to work out why
this should be so, and it is left to the reader. The clip on the connecting
rubber tube is used merely to show that the result is not produced by
simply kinking the tube as the drum is lifted. If the clip is closed close
to the drum, so that the tube remains connected to the soap film but not
the drum, then the film does not blow out at all when the drum is lifted.

In the photograph the drum has been placed in a tea chest and packed
round with straw and paper. This has not been done as a safety precaution but to avoid rapid changes in temperature, which can also cause
the bubble to blow out. A drum brought into the laboratory from a store
room often warms up, and the continued growth of the soap bubble,
which results as the experiment is performed, is annoying. Contact
with the floor can also distort the base of the drum sufficiently to cause
the bubble to blow out, and the chest should be supported on four small
feet at its corners.

Chapter IV

Buoyancy in the Atmosphere

When the atmosphere is still any particular parcel of air neither rises nor falls. Since it possesses weight, its weight must be balanced by the differences in pressure over its surface. Take a block of air with vertical sides and flat horizontal ends. The pressure of the air on the bottom surface will act upwards. It will be greater than the downwards pressure which acts on the upper surface. The difference must support the weight of the parcel of air. If, in place of the air of the parcel, we put another body of the same size and shape, the same atmospheric pressures will act on it as acted on the parcel of air. These pressures just supported the air which the body has displaced. Every body in the atmosphere must, therefore, experience an upwards force equal to the weight of the air it displaces. This accords with the well-known principle of Archimedes, which thus applies to bodies immersed in any kind of fluid, liquid or gaseous.

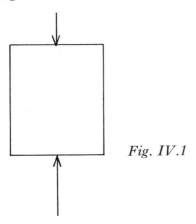

Fig. IV.1

In the case of most solid bodies the weight of the air which they displace is small compared with the weight of the bodies themselves. The buoyancy of the air, therefore, ordinarily passes unnoticed. We cannot swim in the air because we do not obtain sufficient support from it. Nevertheless, there is a force of buoyancy acting on our bodies and our weight appears to be slightly less than it would if we lived in a vacuum. In a vacuum our weight would appear to be about an eight-hundredth greater than it does in air – about 30 g in the case of a 25 kg boy (i.e. about 1 oz for a boy weighing 4 st.).

It is not difficult to demonstrate the existence of the force of buoyancy which the air exerts on all bodies immersed in it. Probably the simplest experiment of all is to blow soap bubbles with town gas. Anything which is lighter than air will float in it, and town gas contains a good deal of hydrogen or methane, and is less dense than air. The buoyancy which a bubble a few centimetres in radius experiences, will support not only the gas but the soap film which surrounds it as well, and the bubble will quickly rise into the air. Those trying the experiment with bottled gas,

Plate IV.1 Buoyancy in the atmosphere. The plastic bag on the left-hand arm of the balance is suspended with its opening downwards. Gas from the mains is liberated gently underneath it. The gas collects in the bag and, since town gas is less dense than air, the force of buoyancy which the atmosphere exerts upon it causes the left-hand arm of the balance to rise.

such as 'calor gas', are doomed to disappointment. These gases are heavier than air and bubbles blown with them fall to the ground in a very determined fashion.

The simple balance used to demonstrate the weight of air can also be used to show the buoyancy of the air acting on a light gas, such as town gas or hydrogen. In Plate IV.1 the plastic bag on the left-hand arm of the balance is open at the bottom and, if one of these gases is released underneath it, it will collect in the bag and the balance of the beam will be disturbed. Similar experiments to those in which a heavy vapour was poured from one vessel into another, can be performed also with light gases, except that light gases must be poured upwards and not downwards. Town gas is the obvious light gas to use for the purpose of demonstration of atmospheric buoyancy, but do not forget that it is inflammable. Do not have flames about when trying the experiment.

In the early days hot air was used in balloons, since hot air is less dense than air at ordinary temperatures. Many hazardous voyages were undertaken in hot air balloons with a fire burning merrily in the basket to maintain the temperature in the balloon itself. There has even been an attempt to revive the process as a sport, using a gas burner to heat the contents of the balloon. Like most other things air expands when

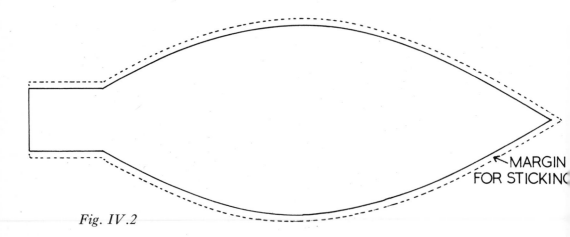

MARGIN
FOR STICKING

Fig. IV.2

heated and thus hot air is less dense than cold. This is, of course, why hot air rises. The force of buoyancy acting on it is greater than its weight.

A hot air balloon can be made from thin tissue paper and sent up safely in the laboratory. Choose a thin paper so as to reduce weight and cut out sections shaped as in Fig. IV.2. Allow 8 mm margins all round to enable the sections to be stuck together. If the edges are then gummed and the sections joined together a spherical balloon with a neck will be formed. A balloon about one metre in diameter should be aimed at. It may be necessary to cut out the sections half at a time and stick them together if large enough sheets of paper cannot be obtained. The neck is essential not only to provide an opening at the bottom to admit the hot air, but also to keep the balloon the right way up when it is floating in the air. If it is dispensed with and just a hole left at the bottom, the balloon will probably turn promptly upside down and empty all its hot air out, and its flight will end ignominiously.

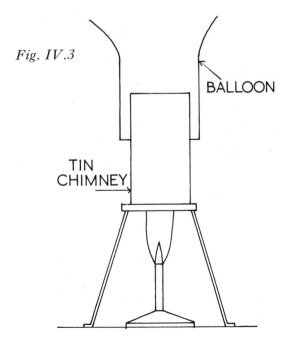

Fig. IV.3

BALLOON

TIN
CHIMNEY

The balloon can be filled with hot air by simply holding it over a bunsen burner. A chimney made by cutting the bottom off a tin should be stood on a tripod, so as to enclose the flame and make it easier to fill the balloon without setting it on fire (Fig. IV.3). It is possible to use an electric fire of the tubular variety to fill it and this is much safer, though the air thus obtained is not so hot and the flight is not so prolonged. The photograph in Plate IV.2 shows a balloon lifting off a 750 W tubular heater, which, in the case of the balloon shown in the photograph, is just sufficient to provide the necessary lift. It all depends upon the weight of the tissue paper which has been used in the construction. However, it takes three or four minutes to fill the balloon in this way and the expectation aroused by the wait renders the final lift-off that much the more dramatic.

Plate IV.2 Hot air is less dense than cool and a balloon filled with hot air will experience a force of buoyancy, which may be sufficient to cause it to rise. The photograph shows a tissue paper balloon, just over a metre in diameter, lifting off a 750-watt tubular electric heater.

4. *Alto-cumulus cloud from above.*
The photograph was taken from a jet
aeroplane flying at a height of about
11,000 metres over central America.
The cloud is at a height of about
5,000 metres.

5. *Typical cumulus, or 'heap'*
cloud.

6. *High stratus cloud. There is no ice halo of 22°*
because there are no ice particles in the cloud. The sun,
however, is surrounded by a diffraction halo.

7. *Low stratus cloud.*

8. *A mist or fog is merely a cloud at ground level. Its depth can be quite small, the tops of trees and high buildings rising above it.*

9. *Nimbus. A view towards the Cuillin Mountains of Skye from near Elgol. Rain obscures the more distant scene. Colour Plate 20, taken from near the same place on a fine day, may be compared with this one.*

*Plate IV.3   The airship R101 at its mooring mast at Cardington, Bed-
fordshire. Filled with hydrogen it met with disaster on its first major long
distance voyage. It set out, rather hurriedly, from its base for India but
crashed at Beauvais in France. It caught fire and most of those on board
lost their lives. (Reproduced by permission of HMSO.)*

Airships designed by Count Zeppelin's engineers, and known by his
name, were used for bombing raids during the First World War. They
caused a good deal of alarm among the civilian population but were
relatively ineffective. Being filled with the very inflammable hydrogen
gas they were easily set on fire and destroyed. After the war and before
the successful development of the large and rapid heavy aeroplane,
much attention was given to the lighter-than-air ship. A very light metal
framework covered with fabric was used to enclose a number of balloons
and the engines were suspended below in gondolas. In the United
States of America helium, which is a non-inflammable light gas, was
used to inflate the balloons, but in England and Germany hydrogen
continued to be used. Although helium is twice as dense as hydrogen
its lifting power is not very different. The buoyancy of a gas depends
upon the difference between its density and that of air. The density of

air, as we have seen, is 1·3 g/l. That of hydrogen is 0·1 g/l, while that of helium is 0·2 g/l (in each case to the nearest tenth of a gramme). Thus the lifting power of hydrogen is 1·2 g/l while that of helium is 1·1 g/l. Except for one or two, all the airships ever constructed ended their careers in disaster. In some cases the metal framework proved too flimsy to withstand the strains imposed upon it by manoeuvring during flight or by differing air currents (the airships did not fly high enough to avoid the weather), many caught fire and there was much loss of life. With the wreck of the British R101 with a high proportion of the country's expert airship builders and operators on board, the project was abandoned and other countries soon followed Britain's example. The sister ship R100, which had made a successful double crossing of the Atlantic to Canada and back, was dismantled. The old R34, perhaps the most successful of the whole generation of British airships, also lived to be finally dismantled in its hangar. Thus ended the era of these large and magnificent leviathans of the air, which joined the dinosaurs and the other extinct animals which failed to cope properly with their environment.

That air is compressible is hardly necessary to demonstrate. It is common knowledge that if the pressure is increased on a gas its volume is diminished, and conversely, if the pressure is diminished the gas expands. Experience with an ordinary bicycle pump is sufficient for these simple facts to be appreciated. Since pressure decreases as one ascends in the atmosphere, it follows that if a parcel of air is lifted by some means to a higher level, coming into a region of lower pressure it will expand. It is for this reason that balloons, which are still sent up into the atmosphere for research purposes and for meteorological soundings, are not filled to capacity at the surface, on launching. They are given only as much gas as will expand to fill them when they have attained the altitude to which they are designed to go.

Expansion leads to cooling. This is the basis for the gradual fall in temperature with height, which is usual in the atmosphere. The rate of fall of temperature is known as the lapse rate. If the temperature of the air is always the same as that which a parcel of gas would assume if raised from the surface to the same height, the rate of fall of temperature with height is called the adiabatic lapse rate. It amounts to about 1° C for every 100 m rise in the case of dry air in the lower atmosphere. The presence of water vapour can modify this very considerably, as we shall see later.

When a parcel of air descends in the atmosphere it goes into regions of higher pressure and is compressed. The opposite happens to what takes place when the parcel ascends. Its temperature rises. This process is also of great importance in the life history of many forms of cloud.

The rate at which the temperature decreases with height in the atmosphere is usually less than the adiabatic lapse rate. An average value is about 0·6° C per 100 metres of ascent, but it is variable from time to time and also it is often not constant for an ascent at one place at a given time. It may have one value near the ground and a different one higher up. The rate at which the temperature drops in the atmosphere with increasing height has an important effect upon its stability. This can be seen from the simple graphs of Fig. IV.4.

Suppose that at a certain height represented by the point O the temperature of the air was 10° C and that the lapse rate was the average, − 0·6° C per 100 metres. At a height of 1,000 metres above O the temperature in the atmosphere would correspond to the point L and be 6° C lower than at O. The temperature of the atmosphere would thus be 4° C. Now let us take a parcel of air from the height corresponding to O and raise it by 1,000 metres. Its temperature would fall at the adiabatic lapse rate and when the air has risen 1,000 metres its temperature will have fallen by 10° C. In other words its temperature would be 0° C. The air which has thus been made to rise would therefore be colder than the air surrounding it. It would be more dense and, unless it was supported in

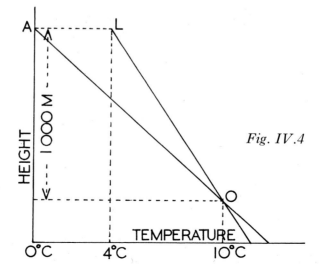

Fig. IV.4

some way, it would fall back again. Similarly, if the parcel of air corresponding to the point O were made to descend, it would warm up by
compression, again at the adiabatic lapse rate, and its temperature
would rise more quickly than that of the air with which it would be
surrounded. Being warmer than the surrounding air it would tend to
rise again. The atmosphere under such conditions would be stable. If
any of it became displaced upwards or downwards, it would tend to
return to its original position.

    If, on the other hand, the lapse rate in the air were greater than the
dry adiabatic lapse rate, the air would be unstable. A parcel of air which
was displaced upwards and cooling at the adiabatic lapse rate would not
cool as quickly as the temperature of the surrounding air diminished
with height, since the surroundings would get colder more quickly.
Being warmer than the air in contact with it, buoyancy would make it
rise still further. The air would be unstable. If some were slightly displaced upwards or downwards, the parcel of air would tend to move still
further away from its starting point.

    Suppose that the lapse rate in the atmosphere on a particular occasion
were 0·6° C per 100 metres, and represented by the line PQ in Fig. IV.5.

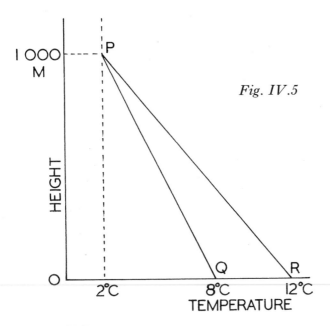

Fig. IV.5

The point Q lies on the surface of the earth, where the temperature of the air is 8° C. Now let us suppose that the sun starts to shine and some air in contact with the surface has its temperature raised from 8° C to 12° C. It will then be less dense than the air which surrounds it and will start to rise. As it does so its temperature will fall according to the adiabatic lapse rate of 1° per 100 metres ascent. At a height of 1,000 metres, corresponding to the point P, its temperature will have fallen by 10° C and it will consequently be 2° C. The temperature of the surrounding air will have decreased by 6° C and in consequence it also will be 2° C. The parcel of warm air which has risen will find itself at the same temperature as its surroundings. The buoyancy of the surrounding air will be just sufficient to support its weight and its tendency to rise will disappear. It will come into a new position of equilibrium at a height of 1,000 metres. In actual fact, of course, the rising air will possess momentum which will carry it somewhat beyond the position of equilibrium, and it will only come to rest after oscillating about this position.

We have assumed in all this that the air is dry, so that no condensation of water vapour takes place, but we shall see, a little later on, that, if condensation does take place, this scheme of things can be altered in important respects. The consideration of this, however, we will postpone until the next chapter.

The temperature of the atmosphere does not, of course, always fall off steadily with height, as we have assumed for the sake of simplicity. Occasionally the temperature may even rise with increase in height over a certain range. Such a reversal of the usual lapse rate is known as an inversion. Inversions often do not extend very deeply in the air, but when they occur they produce layers of great stability. If a parcel of air which is practically in equilibrium to start with, rises, it finds itself soon in contact with air which is warmer than at the start. It soon loses its buoyancy and it will drop back again. Inversions may occur through the over-running of cold air by warmer air above it. They can occur through contact of the lower layers with a cold surface. Inversions near the surface are the rule in polar regions, where the surface is intensely cold and they commonly extend there to a height of about a kilometre. Inversions also form in temperate latitudes, particularly on clear, cold nights when the earth's surface is rapidly cooled by radiation. Warmer air rising in the atmosphere experiences difficulty in penetrating an inversion because the buoyancy drops off and clouds tend to spread out underneath it. Inversions, therefore, are often marked by mist or fog at ground level or by stratus clouds higher up.

In temperate latitudes the average lapse rate of about 0·6° C per 100 metres ascent extends up to heights of about 10,000 metres, rather higher in summer than in winter. At this height the temperature has fallen to about − 50° C to − 60° C and an abrupt change occurs. Above it the temperature no longer falls with increase of height, but remains fairly constant. It even rises again slowly. The lower region of the atmosphere, where the temperature falls off with increase in height, is known as the troposphere. The upper region, where the temperature is more or less constant, is known as the stratosphere. The dividing surface between the two is known as the tropopause. The troposphere is deeper over the tropics than it is over the poles. The temperature of the surface of the earth is, of course, higher in the tropics than near the poles but, because the temperature of the atmosphere continues to fall for a greater height over the tropics, the upper air may be actually colder there than it is over the poles. The tropics are, in fact, warmer up to heights of 10,000 metres, but at greater heights the polar regions are warmer. At greater heights still the differences are smoothed out and the temperature becomes the same everywhere.

The troposphere contains the great mass of the atmosphere, and the weather, as we know it, is produced there. Jet aeroplanes, which fly at heights of about 10,000 metres or rather more, fly in the lower part of the stratosphere, and are, for the most part, above the weather. The storms and nearly all the clouds are formed at lower levels and, once at its cruising height, the jet aeroplane usually finds itself in stable conditions. This factor is almost as important as the great speed of the jet aeroplane in contributing to the success of this form of travelling.

There are four principal processes by which the temperature of a mass of air can be changed. It can gain or lose heat directly by radiation, it can mix with another mass of air of different temperature, it can be heated or cooled by coming into contact with a hot or cold surface and it can have its temperature altered by rising or falling within the atmosphere. Of the four the last is by far the most important. It is responsible for nearly all the rain, hail and snow which falls on the earth, and it is the main factor in the production of clouds. However, all four processes actually occur in the atmosphere and have an effect on the production of clouds and the weather in general as a consequence.

Chapter V

Condensation and Convection

Convection is the name employed to describe the process in which heat is conveyed from one place to another by the actual movement of hot material. It is of great importance in the atmosphere. There are two other processes for the movement of heat, namely conduction and radiation. Heat will flow along a metal bar heated at one end, without the material of the bar moving. This process is called conduction and is of very little importance in the atmosphere. The air is a very bad conductor. Radiation is the process in which heat behaves like invisible light. It is the process by which we can feel the heat from an electric lamp immediately it is switched on. It is the process by which heat can cross a vacuum and it is by radiation that we receive heat from the sun. The atmosphere both receives and loses heat by radiation but we will first look at the process of convection, which is responsible for a good many of the clouds which we see.

Fig. V.1

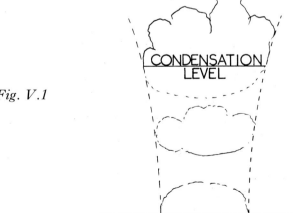

*Plate V.1   A simple cell for the projection of convection currents in water,
by means of a lantern. A lantern giving the image the right way up is
essential, but one can easily be improvised.*

*The cell consists of two sheets of glass held apart by a piece of rubber
tubing and clamped together between two pieces of wood. A small heater is
made from a coil of florists' iron wire and attached to flex. By means of a
glass tube drawn to a very fine point, water coloured light pink by potassium
permanganate is introduced into the bottom, where it forms a dark layer.
On connecting the heater to a cell from a torch battery, it warms the water
in contact with it, causing convection currents to rise.*

Liquids and gases expand when they are heated, and as we have seen,
a parcel of hot fluid rises because the force of buoyancy exerted by
the surrounding fluid is greater than the weight of the hot fluid itself.
The rising body of hot fluid carries heat with it and so the process of rising
currents of hot air is known as convection in the atmosphere. It is well
worthwhile studying the effect of convection in water and a simple cell
for its demonstration is quite easy to make. Plate V.1 is a photograph of
one which is suitable for projection in a lantern. It is of considerable
importance to use a flat cell so that convection currents are limited more
or less to one plane and what is happening is not obscured by currents
moving about in front. Details of its method of construction are set out
in Chapter XII on page 131. It consists of two glass plates separated by
a rubber tube and clamped together between frames of wood. The cell

is nearly filled with water and then, by means of a glass tube drawn out to a very fine point, some water coloured pink with potassium permanganate, is carefully introduced into the bottom. (It may be an advantage to cover the end of the tube with cotton wool to prevent the coloured water squirting out too quickly and stirring up the contents of the cell, thereby making observation difficult.) The coil of wire is a heater made from iron wire attached to flexible leads. The heater is warmed by connecting the flex to a cell from a torch battery. Plate V.2 shows a sequence of four stages as the heated coloured water rises. If projected on to a screen by a lantern which gives the image the right way up, the effects can be very beautiful. Details of the construction of such a lantern are also given in Chapter XII. The first stages are reminiscent of the form of a cumulus cloud, and the subsequent ones, where the warm water spreads out below the surface, of the anvil top of a thunderstorm (see Chapter VI). The descending cool coloured water, heavier than its surroundings on account of the potassium permanganate dissolved in it, shows the baggy form of cumulus mammatus (See Chapter VI p. 74 and Colour Plates 19 and 20).

In the case of the atmosphere, the air is heated by the sun but, all the same, it is heated, in the main, from the bottom and not from the top. Very fortunately for us the heat rays from the sun pass through the air with little hindrance. The air, in other words, is transparent to these heat rays, which are thus able to heat the surface of the earth. Let us consider what happens on a clear calm day in summer. Some parts of the surface of the earth, such as bare rock, tarmac, a town or a field of ripe corn, may get hotter more quickly than the surroundings and the air in contact with them will be warmed. After a while a bubble of hot air may leave the surface and rise. As it does so it will go into the regions of lower pressure, expand and cool. If the hot air rises sufficiently high into the atmosphere, its temperature may fall to its dew point. When this happens the water vapour it contains will condense into tiny droplets of liquid water, and a cloud will be formed. The small light clouds which form during the morning on a summer day, for example, are formed in this way. Their typical shape is a flat base, on which the cloud appears to rest, and an undulating top. The undulating top is simply the top of the ascending air. The flat base forms at the height to which air, warmed at the surface sufficiently to rise, has to be lifted for its temperature to fall to the dew point. Such a typical shape is seen in Colour Plate 11.

As a rule the cloud does not rise into air which is completely still, and usually the wind blows more strongly higher up than lower down. The result is that the top of the cloud is carried along faster than the bottom, so that the cloud slopes in the direction of the wind. This is noticeable in Colour Plate 11 and seen more clearly in Colour Plate 14.

Plate V.2 (Opposite)   A series of patterns formed by the convection currents photographed in the cell of Plate V.1. At first a column rises upwards. The shape is somewhat reminiscent of the rising columns in cumulus clouds but the analogy must not be pressed too far because of the difference in scale. There is a tendency for the descending cooler water to be drawn into the ascending stream, thus illustrating the vortex ring circulation described in Chapter V.

The warm water spreads out near the surface, like the clouds under an inversion or the anvil of a thunderstorm under the stratosphere. When it is cool the coloured water is heavier than its surroundings, because of the potassium permanganate dissolved in it. It descends and in doing so it shows the bulbous form like cumulus mammatus. (Colour Plate 19).

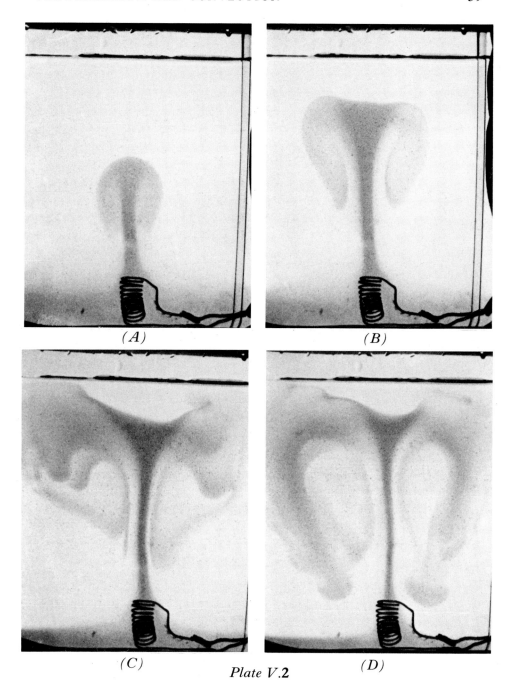

Plate V.2

So far we have only looked at the process of convection in the atmosphere up to the point at which condensation occurs. Below that point the air behaves in much the same way whether it is damp or dry. As soon as condensation occurs, however, a totally different state of affairs arises. This is because of the effect of latent heat.

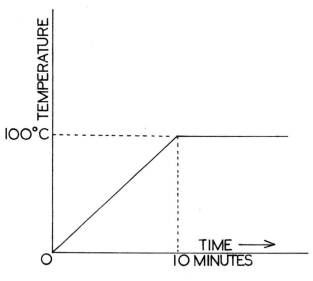

Fig. V.2

If we place a bunsen burner under, say, 1,000 grammes of water and note the temperature at equal intervals of time and plot the results, we obtain a graph like that in Fig. V.2. According to the graph the temperature of the water rises 100° C in 10 minutes, so that in the first part of the experiment, while the temperature of the water is rising steadily, the water must be receiving 10,000 calories of heat a minute. As soon as the temperature of the water reaches 100° C, however, no further rise takes place. The water must still be receiving 10,000 calories a minute although its temperature does not rise. The heat is used to change the water from a liquid to a vapour. This heat is given out again when the vapour condenses back to the liquid. It is known as the latent heat of vaporisation.

Evaporation can take place from water, of course, which is not at its boiling point. Latent heat has still to be used, however, to change the liquid into the vapour. That is why damp clothes feel cold. The amount of the latent heat which has to be provided varies with the temperature at which the evaporation takes place. Watt, the engineer, thought that what he called the total heat of steam remained constant. By the term total heat of steam he meant the amount of heat which had to be provided to convert one gramme of water at $0°$ C into vapour at any temperature $t°$ C. If this was so the latent heat must be greater at the lower temperatures, since less heat would be required to heat the water while still liquid and more would be left for the conversion into vapour. Watt's idea of the total heat of steam being constant proved to be incorrect, but all the same the latent heat of vaporisation is greater at lower temperatures than higher ones. At the normal boiling-point of water at $100°$ C it requires 537 calories to evaporate one gramme of water. At $15°$ C the latent heat of vaporisation of water is 589 calories per gramme. At temperatures around $0°$ C we can take a round figure of about 600 calories per gramme to convert water from the liquid to the gaseous form.

This is a very large quantity of heat to be associated with such a small quantity of water as 1 gramme. It would raise 6 grammes of water from $0°$ C to boiling-point or raise a gramme of copper to the temperature of the surface of the sun, if the copper could remain solid and not itself evaporate. It is not altogether surprising, therefore, that although the amount of water vapour carried in a litre of air is small, the latent heat given out when it condenses has a marked effect upon the behaviour of air rising in the atmosphere. Let us consider air, the dew point of which is $0°$ C. A litre will contain about $0·0029$ grammes of water vapour. The latent heat of vaporisation of this amount of water would be about $1·7$ calories, which would raise the temperature of a litre of air by $5·6°$ C.

Let us get quite clear how the latent heat given out by water vapour when it condenses can raise the temperature of the air in which it is held. Condensation will only occur if the air is cooled, so that we must not look for any actual rise in temperature to take place. What can happen is this. If a process which would lead to a cooling of the air, such as rising and expansion, is taking place, then, when the water vapour starts to condense, the rate of fall of temperature which actually occurs would be less than what it would have been had there been no condensation.

According to how damp it is air will hold more or less water vapour. Not all of what it contains will condense immediately a cloud forms. Once a cloud begins to form the rate of cooling which took place before will be markedly decreased. For rising air the initial rate of decrease of temperature was near to the dry adiabatic lapse rate of 1° C per 100 metres of ascent. After condensation has commenced this may well be reduced to values which are less than the actual lapse rate in the atmosphere, which on an average is about 0·6° C per 100 metres of ascent. When this occurs the rising air remains at a higher temperature than its surroundings. Instead of coming into equilibrium very soon, the up-current is given a new lease of life and it can rise to great heights. Eventually, however, nearly all the water vapour will have condensed and this source of additional energy will disappear. The process is illustrated diagrammatically in the graphs of Fig. V.3.

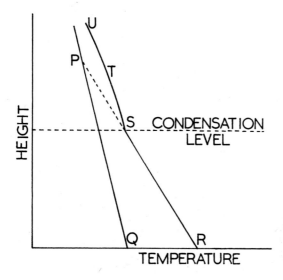

Fig. V.3

QP is the graph of the lapse rate in the atmosphere. The point R represents the temperature of the air which has been heated at the surface of the ground as before. The figure is the same as Fig. IV.5, except that condensation takes place at a certain height corresponding to the point S, so that the point of equilibrium, P, is never reached. After condensation begins the temperature of the rising air follows the more sloping curve ST. As the water vapour condenses the curve ST will bend back again so that its slope decreases once more, and it may finally cut the line QP at some height above P such as U. If it does, the rising air would once again come into equilibrium with its surroundings, and the up-current would cease, possibly after a preliminary overshooting of the mark as the result of upward momentum.

It is of some importance not to fasten our attention too closely on the up-currents, though these up-currents are the cause of the cumulus clouds we are studying at the moment. Down-currents are equally important. When a mass of air descends through the atmosphere, it undergoes the opposite series of events to those experienced by air moving upwards. It descends into regions of higher pressure, it is compressed and warmed. If it contained no cloud, it would warm up at the dry adiabatic lapse rate of 1° C per 100 metres of descent. After a while it would find itself at the same temperature as its surroundings and its descent would cease. If, however, the descending air contained cloud some of the water droplets would evaporate as the air grew warmer, so as to keep the air saturated. In doing so the latent heat would be absorbed. As a result the rise in temperature of the descending air would be less than what it would have been in the absence of cloud. This would aid the descent. Thus the water which air contains, either as vapour or water drops, helps both up-currents and down-currents. In other words it makes the atmosphere less stable. Once started, both up-currents and down-currents tend to go on longer, and the change in height is greater, than would be the case in dry air.

One can learn a lot about clouds by taking a deck chair into the garden, no matter where it may be so long as it has a view of the sky, on a good summer's day when the thermals are giving rise to fairly vigorous cumulus clouds. Set the chair in a fairly low position so that you see the clouds in comfort, and simply watch their development and decay. It is necessary to keep before the mind that what is to be seen is happening on a large scale. The visible sky may well cover an area comparable to that of an English county, any cloud being probably many acres in extent, so that the changes to be observed will take place comparatively slowly, even though the actual velocities of the air currents may be quite large. One must be prepared to follow an individual cloud for five or ten minutes, and for that reason a day on which there is a strong wind is not so good, because the clouds will sail by too quickly.

It is necessary also to be prepared for some initial disappointment, for nearly all the conspicuous towers, on which we naturally fasten our attention, will be found to be decaying. To see the growth of a tower needs careful searching. This indeed is, perhaps, one of the most useful observations to be made, as it throws a good deal of light on the circulation in the cloud. What we see when we look at a cloud is its outside, and it is not here that the up-currents exist. On the outside the air currents are downwards, not upwards. A moment's thought will convince us that it must be so. When a bubble of hot air rises air must flow in underneath

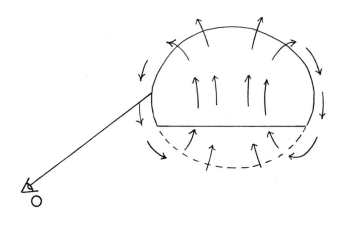

Fig. V.4

10. *A rainbow can only be formed in water drops and is usually only to be seen in comparatively large ones falling as rain beneath a nimbus cloud. The primary bow shows the well-known colours of the spectrum, with the red on the outside. Another faint bow called the secondary, occasionally is to be seen lying outside the primary. It has the red on the inside. Within the primary bow and close to it are sometimes to be seen faint bows, usually of pale purples and greens. They are known as supernumerary bows. They are, however, too faint to reproduce in this picture.*

11. *A photograph of cumulus cloud taken at 10.30 a.m., showing considerable vertical development. Within a little over an hour heavy showers were falling. Early vertical development is a sign of the likelihood of showers later in the day.*

12. *A contrasting picture to Colour Plate 11. The photograph is of cumulus clouds taken on a summer afternoon. The weather was settled and the clouds show very little vertical development.*

13. *Steam is invisible. The cloud coming out of the old steam locomotive is not steam. It is partly smoke from the burning coal, but much of it is composed of a cloud of water droplets which have condensed from the steam when it came into contact with the cooler air outside.*

14. *Rising currents of warm air, which give rise to cumulus clouds, usually meet with stronger winds as they ascend. The result is that the columns of the clouds which they produce are tilted forward in the direction of the wind.*

15. *A sequence of photographs showing the changes in a bank of cumulus clouds, taken at intervals of about 90 seconds. The central column starts to descend and decay immediately and has disappeared by the end of the sequence. The right-hand column develops a little at first, but it too then decays. The left-hand column develops throughout the sequence but shows signs of the onset of decay in the last frame.*

it to take its place and air above it must be moved out of the way to make room for it, into which it can rise. Thus the general circulation associated with a cumulus cloud must be similar to that sketched in Fig. V.4. In the part of the cloud visible to the eye of the observer O, the currents are downwards, the air is being compressed and warmed, and the water droplets are evaporating. The up-currents are situated right in the centre of the cloud, and it is only occasionally that we get a glimpse of them. This type of circulation can be seen in the experiment with the coloured water in Plate V.2 (A). The descending, cooler, coloured water shows a tendency to be drawn into the rising column as in Fig. V.4.

Something resembling the circulation in a cumulus cloud can be seen in the smoke billowing from a garden bonfire, though there are important differences. The hot air containing the smoke arising from the fire forms a more or less continuous column. Air enters at the fire itself so that there is no tendency for that adjacent to the column of smoke to travel downwards, to take the place of the air moving upwards. However, the edge of the column is composed of a series of eddies, the outsides of which move upwards or downwards hardly at all. Nevertheless the general circulation is otherwise similar to that to be expected in the cumulus cloud.

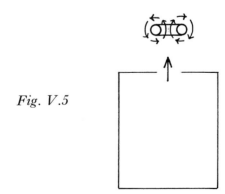

Fig. V.5

There is also a resemblance to the circulation in a vortex ring. Take a cardboard box and cut a hole about two or three centimetres in diameter at one end. Fill it with smoke from smouldering brown paper. The lid should be made of some fairly flexible material and, if this is flicked by the finger or with a stick, a vortex ring in smoke can be made to emerge from the opening. The circulation in the ring is shown by the smoke and can easily be seen, especially after the ring has lost most of its forward velocity. To be analogous to the circulation in a cloud the vortex ring must be sent up vertically. There will then be an up-current in the centre and down-currents on the outside.

Those who possess an 8-millimetre ciné camera, with which it is possible to take single shots, can make a very valuable study of cloud movement and development. The camera must be mounted on a tripod, so that it always points in the same direction, and it is best to use a cable release to avoid vibration when making the exposure. Quite inexpensive cameras will give excellent results. To obtain a good time-lapse film of growing cumulus clouds, one exposure every three seconds or so will be found to be suitable for clouds in the middle distance. When the cloud is near we only look up at the underside, which is not very interesting, and when too distant, the movement appears too small to show up well. Also it is best to choose an occasion, at least for the first time or two, when an individual cloud can be kept under observation for seven to ten minutes, giving 150 to 200 pictures. Later studies under conditions of stronger winds can give very interesting information on the shearing of clouds and turbulence. When the film has been processed (short pieces in black and white are easy to process at home), each sequence should be cut out and the end joined to the beginning to make a loop. If then projected at the normal speed, the movement of the clouds will be speeded up about five hundred times and will be rendered much more obvious. One important advantage of the loop is that it can be left to go round several times in the projector and the opportunity taken to follow various parts of the cloud.

The amount of general 'boil' going on in the atmosphere, which the film will show up, is most revealing. Colour Plate 15 shows a sequence taken of cumulus development, though the stills which are reproduced give little idea of the animation a ciné film will show. The interval between the shots reproduced is about a minute and a half. The initial picture shows the cloud to consist of three main towers, of which the centre one starts to decay immediately. It joins the down-current of air on the outside, and by the end of the sequence, has disappeared altogether. The tower on the left goes on developing throughout, but in the last shot it begins to fall over and will then decay like the middle one. The right-hand tower shows up-currents emerging in the centre. They appear small in comparison with the rest of the mass of cloud but one has to remember that it is only the tips which are seen. The up-current which first emerges soon decays again and it is followed by another in the last shot. This is typical and will be found on many of the time-lapse films taken as suggested above.

The reason why the towers formed in cumulus cloud decay again so quickly can be understood from the considerations into which we have already gone. A bubble of air of radius 100 metres, rising from the surface of the ground, will weigh about 5,000 tonnes, and when moving it possesses considerable momentum. When it reaches the height at which it would be in equilibrium with the surrounding air, it may still be moving upwards with a considerable velocity. Its momentum will carry it upwards above the equilibrium level and, when it has spent its upwards momentum, it will sink back again towards the position of equilibrium. This will start it on its downwards track and, if it has lost a little of its heat by mixing with the cold surrounding air, the evaporation of the cloud as the air subsides will lead to the descent continuing.

Chapter VI

Freezing

When water is cooled to $0°$ C it usually starts to freeze, and when the temperature rises above $0°$ C the ice which has formed melts again. Just as there is a latent heat to be supplied if liquid water is to be converted into vapour, so also latent heat has to be supplied to convert solid water – i.e. the ice – to the liquid form. If a bunsen burner is placed under a tin of melting ice, the temperature of the contents of the can will remain at $0°C$ until all the ice has melted, in spite of the fact that the bunsen burner is supplying heat all the time. The case is similar to the heating of boiling water. There is no change of temperature, all the heat supplied being used to convert the water from one state to another.

The latent heat which must be supplied to convert ice to water is known as the latent heat of fusion of ice. Its value is 80 calories per gramme, and although it is only about a seventh of the latent heat of vaporisation of water, it is still very considerable all the same. It would be sufficient to raise one gramme of copper to red heat. Its effect on the life history of a cumulus cloud can be very dramatic.

Although the freezing point of water is $0°$ C, freezing does not always take place as soon as the temperature of water is lowered to this level. What usually happens is that the temperature of the water falls a little below $0°$ C before freezing commences. Ice then forms and sufficient is produced for the latent heat of fusion – which is given out by the water in this case, as it turns to ice – to bring the temperature of the mixture back to $0°$ C. As further heat is taken away more ice forms, the temperature remaining at $0°$ C until all the water has frozen. Further cooling will then lower the temperature of the ice below $0°$ C. The fall in temperature below $0°$ C which takes place at first, before ice forms at all, is known as supercooling. The amount by which the water in a pond can be supercooled is very small. Ice forms almost immediately the temperature falls to $0°$ C. However, the droplets of water in a cloud can be supercooled very considerably indeed. Some very small droplets can fall to temperatures as low as $-40°$ C before freezing commences. Generally speaking, the smaller the droplet the lower can be its temperature before it starts to freeze. Supercooling in clouds below the freezing point of water is the rule rather than the exception, when the cloud as a whole is cooled below 0° C.

If the supercooled droplets strike an aircraft and break up upon its wings to form a layer of moisture, they freeze immediately. When flying through supercooled clouds an aircraft can quickly accumulate a large quantity of ice, easily sufficient to change its flying characteristics very much for the worse. The control surfaces may be affected so that the pilot loses control and the lift may be so much reduced that it can no longer support the plane. Cooling by expansion, which takes place in the case of the inlet gases to the engine between the carburettor and the cylinder, may lead to ice blockage in the fuel supply. Many serious accidents have been caused in these ways and de-icing equipment has been devised to deal with the menace which icing can be. However, it is still desirable to avoid flying into clouds where icing conditions are severe. It is one of the jobs of the meteorologist to warn pilots of this hazard.

If a cumulus cloud goes on rising, eventually it will reach a height at which its water droplets freeze in spite of the supercooling. In freezing, the latent heat of fusion is liberated. In the case which we have already considered of the condensation of water vapour, the liberation of the latent heat of vaporisation when the water vapour condenses does not actually raise the temperature of the air in which the cloud is forming, although it prevents the temperature falling as rapidly as it otherwise would do as the air rises further. Only a certain amount of the water vapour present condenses at any one instant. If the temperature were to rise again some of the condensed water would evaporate once more. When a cloud freezes, on the other hand, the water droplets are already well below 0° C and a rise in temperature would not cause them to melt again, unless it took the temperature above the freezing point of 0° C. The temperature of the droplets will rise when they freeze because of the liberation of the latent heat, and it is thus possible for the temperature of a cloud actually to rise when freezing occurs. The freezing can also occur rapidly and the effects be very striking.

The liberation of the latent heat of fusion when the droplets of a cloud freeze gives the up-currents of the cloud a new lease of life. Cumulus clouds which have soared to a height where their droplets freeze, suddenly sprout curious growths out of their tops, known as anvils. They are so called because they often seem to be shaped like the

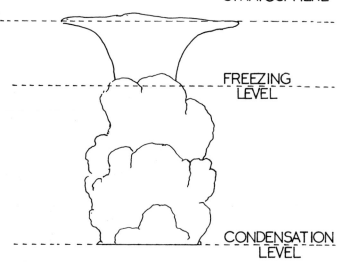

STRATOSPHERE

FREEZING
LEVEL

CONDENSATION   *Fig. VI.1*
LEVEL

anvil of a blacksmith, when seen in the distance. They look very like
cirrus cloud, which is perhaps not altogether surprising since they are
made of the same stuff, namely ice crystals. They are, indeed, sometimes
known by the name of false cirrus.

The anvil-like shape is caused by the great height at which the whole
process occurs. The up-currents containing the ice crystal cloud rise out
of the cumulus cloud of water drops, until they reach the stratosphere.
They cannot penetrate very far into this because their temperatures
continue to fall as the up-rise continues, whereas the temperature of the
stratosphere remains nearly the same at all heights. The rising air
quickly finds itself surrounded by air which is warmer than it is itself,
it loses buoyancy and so spreads out along the tropopause. Seen in the
distance this produces a flat top and the curving sides make it look
like an anvil. When seen from closer quarters so that we have to look up
at it, the anvil looks more like a funnel with a rounded outer edge.
Ultimately it may cover the whole or a very large part of the sky with
cirro-stratus.

Colour Plates 16 and 17 are views of such distant anvils. The one in
Colour Plate 16 was taken from Colombo, Ceylon, over the Indian
Ocean; that in Colour Plate 17 was taken on Newmarket Heath. Both

show the typical anvil shape. The anvil over Newmarket Heath occurred early in the year when both the freezing level in the cloud and the tropopause are lower than in the summer. Near the equator the tropopause is higher than in temperate latitudes even in summer, with the result that anvils occur in the tropics at greater heights. Colour Plate 18 is of an anvil associated with a storm in the French Alps, which is sheared to one side by a wind gradient.

Anvils are only produced in vigorous cumulus clouds and are a sure sign of storms. In the spring they may mean nothing more than an isolated heavy shower, but in the summer they often indicate the presence of thunderstorms. In any case, if anvils are to be seen in the sky in any direction, when setting out for a country walk take a waterproof with you and look out for squalls. Almost invariably rain, usually heavy, will fall from any cloud on which an anvil head has grown and icing seems to play an important part in the mechanism producing lightning.

There was a theory which was put forward in 1935, known after its propounders as the Bergeron-Findeisen theory, which maintained that rain, hail or snow which falls from clouds is caused by the formation of ice as a first step, and that clouds containing no ice particles do not produce anything in the way of rain. This theory is now known to be incorrect, rain having been often observed to fall from clouds the tops of which were above $0°$ C in temperature and which could not possibly, therefore, contain any ice. It is nevertheless true that the icing of a cloud is almost invariably associated with some form of precipitation – rain, hail or snow – so that, at least, it must often be a very important factor in the formation of rain.

Ordinarily clouds, like the cumulus clouds of fine weather, without any ice particles in them, do not produce rain. For the larger drops to collect enough smaller ones as they fall through a cloud and eventually be able to reach the ground, it is necessary for the cloud to possess considerable thickness. It is also necessary for a sufficient number of comparatively large drops to be present to start with, so that by falling through the smaller drops, which have a slower rate of descent, they can grow into raindrops. These conditions appear not to be met with in fine weather cumulus cloud. Such clouds are said to be colloidally stable. The smaller drops do not collect together to form large ones. It is probable that these clouds are not deep enough, and do not last long enough, for drops to grow sufficiently by collision to form rain. Distances of fall of the order of two and a half kilometres and times of about one hour appear to be necessary for the process to be effective.

It was to overcome the difficulty of finding a method to explain how a sufficient number of large drops could be formed in rain clouds, that the Bergeron-Findeisen theory was put forward. The way its authors thought it might happen was as follows. The pressure of water vapour in the air which saturates it in the presence of ice is lower than is the case when supercooled water alone is present, at the same temperature. Air can therefore be saturated with regard to ice while being unsaturated with regard to water. In other words, in such air drops of liquid water will evaporate, while crystals of ice would grow by the condensation of the water vapour on them. If we consider a cloud of water drops in which a few have solidified and turned to ice, the ice particles will find themselves in air which is, for them, supersaturated and they will grow at the expense of the water drops. As the ice particles abstract water vapour from the air the water drops will evaporate to maintain the supply. In this way large particles, which would fall through the cloud sufficiently rapidly to collect smaller drops by contact with them on the way down, might be formed. They would finally fall out of the bottom of the cloud as rain if they melted in the lower and warmer air, or as hail or snow if they failed to do so.

Experimental evidence that this method is of importance in the formation of rain can be obtained from the investigations which have been carried out into the possibility of the artificial production of rain. These attempts have all been based upon the idea of 'seeding' the clouds, and thus they are all dependent upon the ice particle theory of rain formation. The origin of the method of seeding is as follows. If a small particle of the solid is introduced into a supercooled liquid, the whole of it will instantly solidify, providing the supercooling is sufficient. Once a start has been made the molecules of the liquid quickly join the ordered array, which forms the crystal of the solid. Solidification, however, seems to require some nucleus to give it a start. Some of these nuclei occur naturally and are effective at various degrees of supercooling. It is thought that the reason that small droplets can be supercooled to a greater degree than larger ones is that the chance of finding a suitable nucleus is so much greater in the larger drop than in the smaller one.

Substances which have a crystalline structure similar to that of ice can also start the process of solidification. Substances which have been tried for the purpose are the iodides of silver, cadmium and lead, all of which are effective in reducing the extent to which water can be supercooled. The idea which underlies the attempts at the artificial production of rain is, therefore, first to find clouds in which the water droplets are super-cooled by a good margin, and then to inject into them particles of the seeding material, usually in the form of smoke from an aeroplane. Such experiments have had limited success, sufficient to encourage the hope that something of value may result in the end although they are not as yet good enough to warrant the method being applied for practical purposes. The success which has been achieved is, however, sufficient to lend support to the view that ice particles can be effective in the process of rain production in nature.

If this view is accepted, the difficulty is then shifted to the problem of how to account for the rain which does undoubtedly fall out of some clouds which are entirely above 0° C throughout. Such clouds cannot be deeper than about 3 kilometres in temperate latitudes, so that the margin available for the growth of raindrops by collision is not great. Most clouds, under such circumstances, could not produce anything much more than a drizzle. Bowen, however, has pointed out that the larger falling drop will be situated in an up-current and, though falling through the air and passing smaller droplets, it may still be carried up-wards in the cloud. If the cloud itself lasted long enough, the drop could then grow sufficiently large finally to overcome the effect of the up-current supporting it, and fall out of the bottom of the cloud as rain. To produce rain in this way the cloud must not only be deep enough for the process to happen but it must not be tilted by wind shear to too great an extent, so that the larger drops fall out of the cloud prematurely.

Anvil tops can be very persistent. Often they are to be seen towering out of a range of distant clouds at sunset, and they can last long after the cumulus cloud itself has dispersed. Such decaying anvils can lead to an overcast sky after a period of thundery weather. They are ice particle clouds and can give rise to haloes and mock suns in the same way as cirro-stratus cloud. In certain types of weather the anvil can form the major part of a storm cloud, even when it is in a vigorous and youthful condition. Very often quite a number of them are to be seen in the sky at the same time.

Just as the emission of the latent heat of fusion of ice when the water droplets of a cloud freeze can lead to the growth of vigorous up-currents, so the absorption of the latent heat of fusion, should the ice melt, can lead to down-currents. The case is exactly the same as that of the condensation and evaporation of water vapour with the accompanying emission or absorption of the latent heat of vaporisation. The presence of water vapour in air lessens its stability, movement either up or down being accentuated. The same effect results from the presence of water droplets which pass the freezing level going upwards or the melting level going downwards. These two levels will not be the same since the liquid drops solidify only at a temperature well below 0° C, whereas they will melt again if their temperature reaches that value. The anvil thus tends to be more stable, as we have just seen, than is the water drop cloud out of which it grew.

Nevertheless the ice particles, on occasion, do melt and when this happens we obtain the converse of the up-current which formed the anvil, in the form of a vigorous down-draught. This is the condition which produces the best examples of the type of cloud known as cumulus mammatus. Great bags of cold, descending, cloud-filled air are formed which bear some resemblance to a cow's udder – hence the name. To show them up well illumination on the underside of the cloud from a sun low in the sky is required. Under these conditions their appearance can be almost as dramatic as that of the original anvil. Colour Plate 19 is a photograph taken under such conditions. Under normal lighting their appearance is less remarkable, and a typical specimen is shown in Colour Plate 20. They are frequently to be seen after a period of rain or storms. Their appearance can be reproduced surprisingly accurately in circulating water, as we have seen in Plate V.2. After the coloured water which has risen on being warmed has cooled again, it will be more dense than the clear water surrounding it. This is so on account of the potassium permanganate dissolved in it – just as salt water is more dense than fresh. The cooled coloured water sinks and in doing so takes on the baggy mammatus form.

Chapter VII

Some Miscellaneous Examples

So far we have examined the uplifting of air by the buoyancy arising from heating. While it is probably the most important way in which air can be lifted up and gives rise to the commonest and most conspicuous of the clouds to be seen in the sky, it is by no means the only origin of uplift in the atmosphere. A very common source of clouds is provided when a wind blows across a chain of hills or mountains. The air is forced

Fig. VII.1

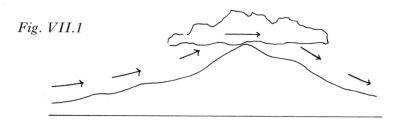

to rise, expand and cool, and we have in this fact the reason why so often mountains have their heads in the clouds. Driving to the mountains from a little distance we may have clear skies and bright sunshine on the way, only to find the summits obscured by cloud when we get there. The wind may well be blowing from the direction from which we have come, but it is no use waiting for the clear weather to arrive before ascending the mountain. The wind blows through the clouds on the mountain top. They form on the windward side, and dissolve again on the leeward side as the air is warmed by compression as it descends. If no water falls out of the cloud on the mountain top the levels of the cloud base will be much the same on either side of the range. If water is lost on the way over, the cloud will be higher to leeward than to windward but the effect can be complicated by eddies formed near the top as the wind sweeps across the summit or ridge.

Cumulus clouds also develop over mountains in fine weather before they do elsewhere. This occurs for a number of reasons. In the first place bare rock, of which the tops of high mountains are usually composed, heats up rapidly in the sun, giving rise to thermals. In the second place these thermals often contain air from lower levels, which may be moist and which, in any case, has already been raised by the wind blowing across the mountains. In the third place the thermals start from high up and have less distance to traverse before reaching the condensation level and thus less opportunity of coming into equilibrium before getting there. Colour Plate 20 shows a group of cumulus clouds over the Cuillin Mountains of Skye, taken from near Elgol.

Another way in which air may be lifted up is by turbulence. When a breeze blows over land, eddies are formed as it blows across trees, hillocks and other obstacles. This results in a general stirring-up of the lower layers of the atmosphere. Many fine days, particularly in summer, start with an overcast of low clouds a hundred or two metres above the surface. It has been suggested that the origin of these clouds is to be found in the turbulence of the lower layers of the air. During the night the surface of the ground cools down, and so does the air in contact with it. Such cooled air has only to be lifted a short distance before its temperature falls to the dew point and clouds form. Turbulence can supply the required lift.

During the morning these low clouds often disperse. In common parlance the sun is said to 'burn the clouds up'. This, of course, is only a figure of speech. Clouds can exist throughout the day in the most brilliant of sunshine. They reflect most of the energy which falls on them. What is more likely to happen is that enough heat radiation penetrates the layer of clouds to warm up the surface and the air in contact with it, so that turbulence is no longer sufficient to raise the air to the necessary height for its temperature to fall to the dew point. That the temperature of the ground rises as the sun gets up even on a cloudy day in summer, can easily be verified by a screened thermometer. A greenhouse equipped with a maximum and minimum thermometer shows the effect very well, since the greenhouse 'traps' the radiation which gets through. The mercury leaves the minimum index soon after sunrise, although the temperature does not rise rapidly until the clouds have dispersed.

We can represent what happens in a somewhat idealised schematic way by the lapse rate graph in Fig. VII.2. The mixing of the lower layers by turbulence will result in a lapse rate approaching the adiabatic,

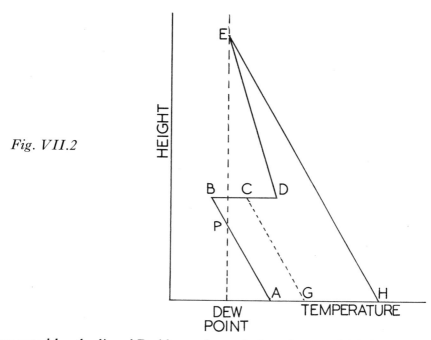

Fig. VII.2

represented by the line AB. Above the turbulent layer, which has been cooled by radiation from the ground during the night, an inversion could develop, as indicated by the line BCD. If AB cuts the line marking the dew point of the lower layers, at some point such as P, clouds will form at this level. When the sun raises the temperature of the ground to G, adiabatic cooling will no longer be sufficient to lower the temperature to the dew point before the rising air comes into equilibrium at C on the inversion. The sky will then clear. It will not be until the temperature of the air near the ground reaches the value represented by H that normal cumulus clouds will form at the height indicated by the point E. There may thus be an interval between the clearance of the low cloud and the formation of the normal summer cumulus.

The rate at which the cloud will clear will depend upon the rate at which the ground warms up. This, in turn, will depend upon the thickness of the cloud cover and on the nature of the ground. Over large areas of wet clay soil the process will be slower than over areas of dry sandy soil. The process also takes longer in winter than in the summer. When the sun is low the clouds may never clear at all throughout the day.

Often the sun rises into a clear sky, but within a short time it clouds over. Here the mixing giving rise to the line AB may be incomplete. Nevertheless, there may be a high lapse rate because of turbulence, and slight warming by the sun may cause the layer to overturn, setting up the conditions illustrated in the figure.

Cumulus clouds are often arranged in long lines or streets. If a particular hot spot on the earth's surface gives rise to a series of rising bubbles of hot air, they will be carried along by the wind and result in a long line of cumulus clouds. Colour Plate 22 shows such a street lying in the direction of the wind. Such streets are often most readily to be seen soon after the cumulus clouds begin to form. Long lines of them can easily arise. Colour Plate 23 is of very long streets formed over southern England. The photograph was taken from a height of about 9,500 metres (about 28,000 ft) over London airport, while in a queue waiting to land. The cloud streets stretch to the South coast. The small dark patch to the right of the reflection of the sun is the Isle of Wight. The distant line of clouds lies over France. The white marks in the central upper part of the picture are reflections in the window of the aeroplane.

We have already seen how the anvil tops of thunderstorms spread out under the stratosphere. Inversions, of course, also occur at lower levels than this and can lead to a similar spreading out of the cloud below them, or a little way within them. In an inversion the temperature rises in the air for a short distance with increase in height. A rising column of warm air thus finds itself in contact with air which becomes steadily warmer as it ascends. The buoyancy provided by the air thus falls off and unless the up-current is a very vigorous one, it will be unable to penetrate the inversion completely and emerge again on the upper side of it. Clouds thus tend to spread out when they encounter an inversion in their path. Indeed, inversions are usually marked by layers of stratiform cloud entrapped within or below them in this way.

Colour Plate 24 shows a group of cumulus clouds which are meeting with an inversion at a considerable height. The inversion is marked by the layer of alto-cumulus cloud at the top of the picture. Above it lies a layer of cirrus cloud. Below it are a number of cumulus clouds. As the peaks of the cumulus clouds reach the inversion they are insufficiently vigorous to penetrate through it and bend over, producing something approaching the anvil shape with which we are already familiar. A similar state of affairs occurs, as we shall see later, when a depression approaches. An inversion forms overhead, the height of which gradually decreases as the depression gets nearer. The cumulus clouds are confined below it and are finally suppressed. The vigorous up-currents of thunderstorms are often able to force their way through moderate inversions, which then often show up against the white columns of the clouds as horizontal dark bands of stratus. One such band is visible above the tower of the château in Colour Plate 18. Inversions, which do not usually extend over more than a few hundred metres, may be as little as a fraction of a degree Centigrade overall, or as much as 6° C or 8° C.

A change in the pattern of the clouds is common at the coast. The clouds in Colour Plate 23, for example, appear to begin at the coast and there is a distant bank of clouds over France. This can be accounted for by the existence of thermals rising from the surface of the ground, the temperature of which rises rapidly in the sunshine, whereas the temperature of the sea away from the immediate vicinity of the sandy beaches is little affected by it. In calm weather the normal pattern of the wind near the coast is that of the 'land and sea breezes'. The sun warms up the land much more quickly than it does the sea, as those who bathe off steep rocky coasts soon find. It is only in stretches of shallow water over a sandy bottom that the sea warms up appreciably. On the other hand, of course, the temperature of the sea in winter is not so vastly different from what it is in summer, which takes some of the glamour off winter bathing in the sea. However, to return to the question of land and sea breezes, a sea breeze will develop in the afternoons in otherwise still conditions in summer, because the warming of the air over the land causes it to rise and be replaced by air from over the sea. The rising air over the land leads to cumulus clouds as soon as the thermals reach the condensation level. Similar effects will also take place whenever there is a wind off the sea. In winter the sea is commonly warmer than the land and the opposite effects can then occur. There is thus often some kind of indication of the coast line in the pattern of the clouds above it.

Sometimes when the lapse rate is high and the air is consequently not very stable, the heating produced by the liberation of the latent heat when water vapour condenses is sufficient to give rise to an up-current, and the development of a cloud of the cumulus type with little or no thermal rising from the ground at all. Of course some initial motion is required to cause a little condensation to trigger the process off, but once set going it will be able to maintain itself by further condensation and the liberation of more latent heat as the upward motion continues. A little turbulence, which is common in air which is not very stable, can serve as the method of triggering. In turbulent air there will be both up- and down-currents and some of the former may lead to condensation if the moisture content is high. Turbulent air gives rise to 'bumpiness' which is noticeable under these conditions by passengers in aeroplanes. The effect of the latent heat diminishes after most of the water vapour has condensed but, all the same, columns of cumulus clouds produced in this way can be quite high. They are commonly also fairly narrow.

Ascending columns of cumulus clouds can arise in this way at all levels in the atmosphere below the stratosphere. When the upper air is unstable alto-cumulus clouds are generated. The normal alto-cumulus cloud, as in Colour Plates 3 and 4, is not very thick, but when the air is unstable there can be considerable vertical development. Vertical development in alto-cumulus cloud gives rise to the variety known as alto-cumulus castellatus. It is not very common and appears more especially in the morning and evening. It is, of course, a sign of instability in the upper atmosphere and it thus tends to occur in a thundery type of weather of which it is often an advance warning.

The last types of cloud which we will discuss in this chapter are those which arise from eddies and waves in the atmosphere. The eddies which are directly responsible for the formation of clouds are those which rotate about a more or less horizontal axis. Cyclones are, of course, nothing more than gigantic eddies which rotate about a vertical axis, and they give rise to enormous cloud systems, but in their case the formation of cloud is an indirect process and they form a class by themselves which we will discuss later. Eddies rotating about horizontal axes arise in the lee of cliffs and mountains, and also between layers of air moving at different speeds. In the latter case they form a kind of roller bearing between the two layers, as in Fig. VII.3. If the air is damp enough cloud may form where the air is lifted by the rotation and it will disappear again where it descends. The type of cloud which such rotation produces is thus composed of long lines of cloud lying across the wind and moving with it at a speed which is intermediate between the speeds of the two layers. An eddy behind a hill (Fig. VII.4) may also lead to the formation of cloud towards the top of the lee slope.

16. Distant thunderstorm anvil over the Indian Ocean, photographed from Colombo.

17. A typical anvil-shaped top taken over Newmarket Heath early in the year. The freezing level is comparatively low and the anvil forms at a low level, and its top, which forms under the stratosphere, is also lower than it would be later in the year. (Photograph by Miss G. Jones.)

18. An anvil in the French Alps, which has been drifted to one side in a wide gradient.

19. *Cumulus mammatus. Great bag-like masses of air cooled by melting ice crystals, descending from an overlying anvil top. The illumination from a low sun shows them up to advantage.*

20. *Cumulus clouds over the Cuillin Mountains of Skye. Cumulus clouds form readily over mountains, partly because the mountains cause the air to rise, partly because they are formed of bare rock which heats up rapidly in the sun, and partly because the warm air rising from them starts at a higher level to begin with.*

21. *The 'table cloth' on Table Mountain. Whenever air is made to rise over a mountain its temperature will fall and, if sufficient, it will lead to the formation of cloud. (Photograph by Mrs E. A. Holland.)*

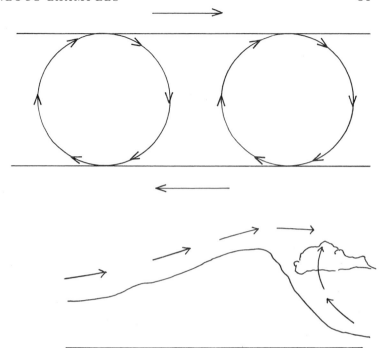

Fig. VII.3

Fig. VII.4

A particularly characteristic form of cloud is produced by waves in the air. For waves to exist there must be some kind of restoring force which can act on air when it is displaced, tending to restore it to its position of equilibrium. In air, as in water, the restoring force can be provided by gravity and this can only act if there are layers of differing densities lying one above the other, like oil on water. Alternatively, air being compressible, the restoring force can derive from pressure differences. When the flow of air is distorted by passing over a range of hills, or moving out over the sea after passing a high cliff, the dividing line can be set oscillating, producing waves on the surface of separation. These waves are like the stern waves formed by a ship ploughing through the sea or the waves behind a boulder in a stream. They maintain station with the obstacle and they are thus stationary relative to an observer on the ground. Cloud may form where the wave carries the air upwards and it will evaporate again where it carries the air downwards. The cloud, being attached to the wave, will also, therefore, appear stationary to an observer on the ground. Like the clouds on the top of a mountain, the wind blows through the clouds, which are continually forming in front and disappearing behind. Such clouds often possess a lenticular shape – a term which simply means having the shape of a lentil, convex on both sides, like a lens. The clouds are known as alto-cumulus lenticularis or cirro-cumulus lenticularis, etc, as the case may be. Colour Plate 25 is a photograph of this type of cloud, taken in the Midlands of England. It is difficult to say exactly what has caused a particular cloud such as the one illustrated, but wave clouds originating in the Welsh hills have been observed over the Midland counties.

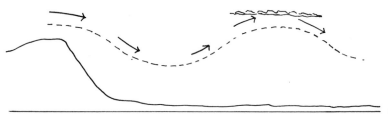

Fig. VII.5

The examples of clouds which have been given in this chapter illustrate the very great importance of the process of cooling by expansion in the atmosphere. It also plays the principal role in the formation of the thick masses of cloud associated with cyclonic depressions which we shall consider in Chapter IX. Mists and fogs, on the other hand, usually owe their origin to a different process which we shall discuss in the next chapter before returning to the question of cyclonic clouds and the process of cooling by expansion which gives rise to them.

Chapter VIII

Fogs and Smogs

We have already come across the process of transference of heat known as radiation. Heat radiation has properties similar to those of light. All bodies emit heat radiation, including those whose temperature is constant. The latter do not cool, because they absorb from other bodies as much heat as they radiate themselves. The whole of space is filled with heat radiation. However, if a body is placed in a stream of strong radiation, such as sunshine, it will absorb more than it emits and its temperature will rise. If it is so placed as not to receive its normal income of radiation from other bodies, it will radiate more than it receives, and its temperature will fall. Cooling of the surface of the earth by radiation and subsequently of the air in contact with it, is one of the most important causes of fog.

During the day the surface of the earth warms up as the result of the absorption of radiation from the sun. During the night it cools because it radiates more energy into space than it receives. Clouds reflect much of the radiation which falls upon them. If the night is cloudy much of the radiation which leaves the earth's surface is reflected back to it by the clouds, and the cooling is much less intense than it is on clear nights when the radiation has a more or less unimpeded path into outer space, from which, of course, it does not return. The only radiation from space which reaches the earth, apart from that from the sun, arises from the stars, the very, very thin interstellar matter, together with the little sunlight reflected by the moon (if visible) or the planets. It amounts to very little in comparison with the energy lost by the earth's surface at night. The cold surface of the earth cools the air in contact with it and, if the cooling is sufficient, condensation of the water vapour in the air will take its place.

The first sign of the process of condensation is often the production of dew, especially in summer. Dew forms on the blades of grass, the leaves of plants and on objects nearby. Nearly saturated air comes out of the stomata – the minute pores on the leaves of plants – and the effect of cooling is naturally felt near to vegetation in the first place. In winter, if the temperature is low enough, the moisture may be deposited as hoar frost rather than dew. If the air is perfectly still the lower layers only of the air can be cooled by conduction of heat to the cold ground, since air is such a bad conductor, and only a very shallow mist or fog could be produced in this way. It has been calculated that, under such circumstances, it is unlikely to be deeper than a little over a metre – say four feet. Very shallow fogs do occur occasionally, such as that in the photograph of Colour Plate 8. It is unusual, even in very calm weather, however, for the air to be perfectly still. Cold air, being denser, tends to drain into hollows and slight turbulence will mix up the lower layers and produce a fog which, although still shallow, is too deep for us to see over the top. Such shallow fogs can be very dense and interfere with traffic. Sometimes the moon or the stars may be visible overhead although visibility along the ground may be very poor. For a fog of any depth to form over flat ground a light wind of about 5 to 6 kilometres per hour (say 3 to 4 miles per hour) is required.

Fogs formed in this way are known as radiation fogs, and a large proportion of those experienced in the British Isles are of this type. In the summer they may form around dawn but as the sun gets up and warms the surface of the earth the resulting circulation of warmer air soon disperses them. The cool surface air is not very thick and above it is usually an inversion. The process of clearing the mist by convection in the surface layer may convert it first of all to a layer of low stratus cloud under the inversion, as described in the last chapter. Further warming of the surface will cause this to disappear also and the remainder of the day may be bright and sunny.

In winter radiation fogs may be very persistent. The warming of the surface may be insufficient to dissipate the fog because so much of the sun's radiation is reflected by the fog itself, and so fails to reach the surface. Frequently no circulation is able to penetrate the inversion overhead. When this happens smoke and other substances from domestic fires and industrial processes, which pollute the air, accumulate in the lower layer of the atmosphere. The highly unpleasant conditions, dangerous to health as well as frustrating to traffic, known by the word 'smog' – a combination of smoke and fog which is exactly what the thing itself is – arise.

Smogs are of two kinds. Those which affect the cities of Britain belong to the group known as sulphur smogs. They owe their unpleasant and dangerous character to the presence of the oxides of sulphur which get into the atmosphere from the burning of coal. These gases accumulate in the air trapped under the inversion and, if such conditions persist for long, can reach dangerous concentrations. The first to be affected are those who suffer from bronchitis and pneumonia. At least three major smog disasters have occurred in recent years. In 1930 a lethal fog settled over the valley of the Meuse in Belgium. In October 1948 a similar thing happened at Donora, near Pittsburgh in the USA, and in December 1952 in London, England. In each case the air in a valley or basin became hemmed in by an inversion which formed a lid on top of it and prevented any ventilation from taking place. The air in all these localities was heavily polluted with the products of burning coal. These included, besides a high proportion of sulphurous gases, some oxides of nitrogen, as well as tar and fine ash. The results in all three cases were disastrous. In London the normal death rate jumped by between 4,000 and 5,000 in the four days during which the smog lasted. Follow-up investigations of the Donora smog showed that many of those who sur- vived an attack of bronchitis during the smog became much more liable to subsequent attacks and their death rate was higher than normal in the area. This again may have been due to the effects of the smog being more severe on those with indifferent health. The institution of 'smokeless zones' in cities has done a great deal to reduce atmospheric pollution, but the sulphur content of the products resulting from burning some smokeless fuels is still high. These gases, though invisible, can still be trapped under an inversion and too much must not be expected of the policy of smokeless zones, valuable as it undoubtedly is. Smokeless zones make a significant contribution towards decreasing the problem but they are invaded by smoke from neighbouring zones. Town gas can be freed of much of the sulphur, and electric power stations can treat their fuel and flue gases. The sulphur content of the gases from the burning of anthracite can, however, still be high. To see the effect of a limited smokeless zone introduced into a city, any large park can be taken as an example. It will be freer from the production of smoke than any smokeless zone can ever be made to be. It is doubtful if any difference would be apparent to the eye although measurements do show a decrease in the pollution.

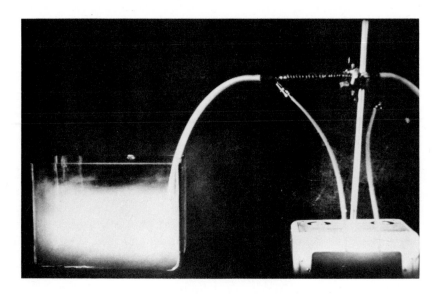

Plate VIII.1   An experiment to illustrate the formation of Los Angeles smog. Ozone is being passed into a tank containing the vapour from a little turpentine. It reacts with it to form a dense white cloud. It is thought that the well-known smog which besets Los Angeles arises from a similar reaction between ozone and unburnt fuel from the exhausts of motor vehicles.

The second kind of smog is typified by that which dogs Los Angeles on about one day in three throughout the year. It too depends upon the development of an inversion. Los Angeles lies on the narrow coastal plane between the sea and a range of mountains in southern California. The rising of heated air from the surface draws in cold moist air from the sea. This cool air cannot escape over the mountains to the east and, in this lake of colder air which accumulates, the atmospheric pollutants are continually concentrated. In the case of the Los Angeles conurbation they come not from the burning of coal but of petroleum, which is, of course, burned in motor vehicles. Los Angeles is, in fact, a conurbation spread over a wide area with a population of some six million. The method of transport upon which the people rely to get to work and home again is the motor car, of which there are about three million. The petroleum is incompletely burned by the motor car engine, the exhaust from which contains a lot of hydrocarbon vapours. The air of Los Angeles also contains ozone which is thought to come from nitrogen dioxide also contained in the exhaust, as a result of a photochemical reaction under the influence of sunlight. Ozone acts with many hydrocarbons to produce a white cloud, as the experiment of passing ozone into a vessel containing the vapour of turpentine shown in Plate VIII.1, demonstrates. A similar state of affairs also arises further north, in the conurbation centred on San Francisco. Here a lake of cold air off the Pacific is also impounded under an inversion in the air by surrounding mountains. These automobile smogs are irritating to the eyes and seriously decrease visibility. The method for reducing their incidence would appear to be to secure a more complete combustion of the fuel in the engine in the motor car. The slow-burning, high octane rating fuels are the worst offenders and 'after burners' to be fitted to the exhaust have been devised to burn up the unconsumed fuel.

Even in the absence of pollution from automobiles, the Californian coast is very liable to persistent fog in summer. Off the shore is the Californian current of cold water from the north which also wells up along the coast from great depths. The interior is largely desert, which heats up rapidly under the sun. The air rises over the interior and cold sea air, often containing sea fog, flows in to replace it. The sea fog is formed in this case by the contact of warmer moist air from farther out over the sea, with the cold current nearer the land.

The system of land and sea breezes can also give rise to fogs on or very close to the coast. If damp warm air is carried from the land over a cooler sea by a gentle wind, a fog may easily develop in it. When the sea breeze develops later in the day because of the heating of the land by the sun, this fog may be carried back over the land. Usually it penetrates only for a short distance inland, disappearing as the air is warmed up again over the land, but at some places, when it is strongly developed, it can penetrate as much as 60 kilometres (40 miles). In still air a continuous circulation may arise, the air rising over the warm land, passing out to sea at a high level, sinking over the sea and returning along the surface, (Fig. VIII.1). If the air is damp, contact with the sea may produce fog which can be carried inland by the circulation.

Sea fogs arise when warm moist air is carried across cold water. The persistent fogs off the Grand Bank of Newfoundland are caused in this way. To the south lies the warm water of the Gulf Stream. Air passing

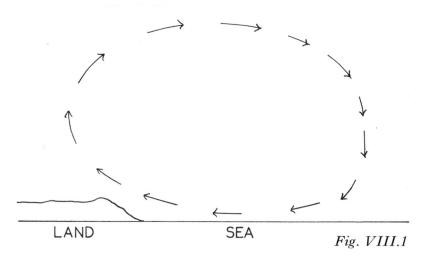

LAND   SEA

Fig. VIII.1

over the waters of the Gulf Stream becomes laden with moisture. To
the north is the cold Labrador Current, which travels down the north
eastern coast of America. When the moist air comes into contact with
the cold water, it is cooled and fog forms. Whereas radiation fogs over
land form as a rule only when the air is still, sea fogs can occur under
much more windy conditions. Sea fogs can be carried inshore by sea
breezes arising from the heating of the land as we have just discussed,
or by means of a more persistent wind. The latter can be of long duration.
The north-east coasts of England and Scotland are often subject to
invasion by sea fog in this way. Because they can form under much more
windy conditions than radiation fogs inland, sea fogs are more common
than are land fogs.

Tropical air being carried to more temperate regions can also lead to
the production of fog over the sea, because of the gradual cooling of the
air as it leaves the tropics. Because there is less turbulence over the sea
than over the land and little differential heating of the surface to cause
thermals, the condensation of the water vapour in the tropical air tends
to take place at the surface of the sea. The flowing of warm moist air over
cold ground in winter can also be a cause of mists and fogs.

Occasionally sea fogs can also be caused by cold air moving over
warmer water. Near the surface of the sea the air may be saturated. As
it mixes with the cold air above, its moisture may condense and produce
fog. The effect can often be seen in canals or ponds used for the dis-

charge of cooling water by power stations or other industrial plants. It is the same as the familiar steaming of the bath water in a cool bathroom. Such fogs are more common in the arctic regions where the sea is warmer than the air above it, and are sometimes known as arctic smoke.

Large masses of air slowly descending, and warming as they do so, can lead to inversions near the ground. Such subsidence occurs in anticyclones, which also tend to be associated with calm conditions. Anticyclones also tend to persist in one area for a long time, during which the general pattern of the weather remains the same. Anticyclones in winter thus often lead to prolonged foggy weather.

What is to some extent the opposite of this, namely the blowing of light winds up a steady slope, can also lead to fog. For this to happen a long gradual slope is required. This is found, for example, in the western central area of the United States of America, where there is a gradual rise to the west. An east wind blowing up this slope will produce foggy weather, the fog in this case arising from the cooling produced by the expansion of the air as it is forced to rise.

During the daytime the flow of air is usually turbulent over the land. Because of heating below, the air is less stable and eddies form in the wind. At night air flows, which are not too rapid, are smooth and follow smooth stream lines. Air cooled by radiation on a hillside at night will flow slowly down it without mixing with the warmer air above it, until it reaches the bottom of the valley. There it will collect and form a lake of cold air, often filled with mist or fog. Villages tend to be built in the 'shelter' of the valley, but the people are here likely to experience more frost and fogs than those who live on the hillside. This fact is of considerable importance to fruit growers, who do not want the blossom to be nipped by frost. High hedges and anything which will dam up the cold air, as it flows down from above, should be avoided at the lower boundaries of orchards. The layer of cold air flowing down a hillside may be quite shallow, although deep enough to catch the blossom on the fruit trees. Fires may be lighted with the object, not so much of actually warming the air of the orchard, as of causing convection currents which will mix the lowest layers with the warmer layers immediately above. Air screws have been tried, to achieve the same objective.

Chapter IX

The Clouds of Cyclones

The pattern of the clouds over Great Britain is very often determined by the large eddies of circulating air which pass over it or nearby. They are responsible for many of the rapid changes which take place from day to day, and almost from hour to hour. They develop out of the different properties possessed by cold polar air on one side, and warm tropical air on the other, which meet along a front out over the ocean. The two air masses travelling at different speeds develop eddies between them along the line of separation – giving the effect of enormous ball bearings, as we have seen in other cases where two streams of air meet. The centre of the eddy is a region of low pressure and the winds blow round this in an anticlockwise direction in the northern hemisphere.

The reason that the winds blow round the area of low pressure and not straight into it so as to destroy it, is connected with the rotation of the earth. If a difference of pressure is created in a fluid which is free to move, it will immediately flow from the regions of higher pressure to those of lower, until the pressure is equalised throughout. For example, if water is poured suddenly into one end of a tank, it will, for a moment, make a hill of water at that end. This will become an area of greater pressure because of the increase in depth of water there, but it will be immediately destroyed because the water will flow away from the hill so as to make the surface horizontal again everywhere in the tank, and thus equalise the pressure.

If the fluid is rotating, however, the conditions are different. When a rotating body contracts its rate of rotation increases. If we swing a small weight on a string round in a circle and allow the string to wrap itself round the finger, the speed at which the weight goes round increases, as it is drawn in towards the centre. If, before pulling the plug out of a bath of water, the water is set rotating slowly, a vortex will form above the outlet as the water is drawn in towards the waste pipe. Everything on the earth's surface takes part in the daily rotation of the earth on its axis. Viewed from the direction of the Pole Star the earth appears to rotate in an anticlockwise direction. If the air travels directly inwards to a centre of low pressure, this rotation will be speeded up. Instead, therefore, of arriving at the centre of low pressure, the air will travel round it, like the water emptying out of the bath.

Any map of the northern hemisphere is drawn from a viewpoint situated above the area mapped and thus in the general direction of the Pole Star. The rotation of a cyclone, which is an eddy round an area of low pressure, is simply the rotation of the earth speeded up. On a map of the northern hemisphere, therefore, the circulation of a cyclone will be shown as anticlockwise. Maps of the southern hemisphere, however, are drawn from a point of view in a direction away from the Pole Star. Cyclones in the southern hemisphere will be shown on maps as rotating in a clockwise direction. In space both rotations are in the same direction. It appears different in the two hemispheres because it is viewed from different sides. If it were to be the habit of people in the southern hemisphere to draw their maps as seen from the direction of the Pole Star the circulation would appear the same as in maps of the northern hemisphere. This, however, would not be very convenient for them, since to read them they would have to hold them above their heads, in order to be able to look in the correct direction. Their maps would then give a 'worm's eye' view of the country.

It is theoretically possible to detect the direction in which the earth rotates by emptying a bath of water which is perfectly still relative to the surface of the earth, and noting the direction in which the vortex over the waste pipe rotates. The rotation of the earth, once in twenty-four hours, is too slow, however, to be detected in this way. It would be impossible to ensure that there was no chance rotation in the water to start with, which was less than the once round in twenty-four hours which we would be looking for.

The result was stated rather differently by Professor Buys-Ballot, in the law which usually bears his name. Buys-Ballot was professor of physics at Utrecht and he stated his law in 1857. It is that when a wind blows from a region of high pressure to one of low, it turns to the right in the northern hemisphere and to the left in the southern. This would, of course, lead to anticlockwise circulation in the northern hemisphere and to clockwise rotation in the southern. Once again this is only a different way of saying that the direction of circulation is the same in both cases but that people in the southern hemisphere are pointing in the opposite direction to those in the northern. Besides the regions of low pressure in cyclones, regions of high pressure also occur in the atmosphere, and these are known as anticyclones. Air tends to flow outwards from these areas. Since it also turns to the right in the northern hemisphere, the circulation round an anticyclone is clockwise in that hemisphere. In the southern hemisphere it is anticlockwise.

Cyclones, being derived from the conflict between cold arctic air and warm tropical air, contain blocks of air derived from these two different air masses. A typical cyclone of temperate latitudes is hundreds of miles across and it is not possible for the two air masses to mix together quickly and thus attain middle values for their temperature and content of water vapour, intermediate between those which the two air masses possessed to start with. Warm and moist tropical air, and cold dry arctic air, become caught up together in the great swirl of the cyclone. The warm air gives rise to a warm sector and the arctic air to a cold sector. Because of the difference in temperature the warm air tends to ride up over the cold air, while the cold air tends to undercut the warm. The general circulation of the two different kinds of air in a cyclone is represented diagrammatically in Fig. IX.1. The area where the warm air is

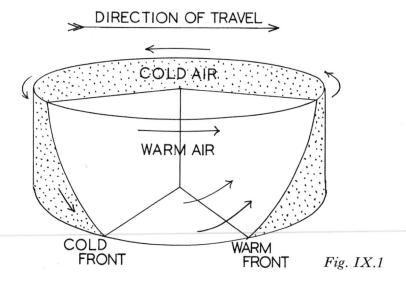

Fig. IX.1

in contact with the surface is known as the warm sector. The divisions between the warm and cold air are known as fronts. The general direction of travel of the cyclones over Great Britain is from south-west to north-east. The leading edge of the warm sector at the surface is known as the warm surface front, or simply the warm front. As it passes any point there is a sudden rise in temperature, corresponding to the sudden change from polar to tropical air. The hind edge of the warm sector is known as the cold front and it marks the passage to the colder and drier air.

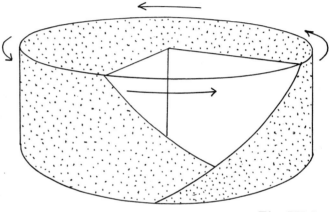

Fig. IX.2

As the warm air rises over the cold air in front of it, and the cold air undercuts it behind, the area of the ground in contact with the warm air continually diminishes, and finally disappears. When all the warm air has left the surface, the cyclone is said to be occluded (Fig. IX.2). As a rule the warm air is in the south or south-western part of a cyclone. On a map showing the isobars (lines of equal pressure) a depression with a well-marked front leading the warm sector and a cold front closing it from behind appears, as in Fig. IX.3. The wind blows practically along the isobars. Fig. IX.4 shows the appearance of a depression in which the warm sector has been partially occluded. On weather maps a cold front is marked by a line of spikes, a warm front by a line of rounded blobs, while an occluded front is marked by the two signs arranged alternately. The signs are placed on the line representing the front, on the side towards which it is moving.

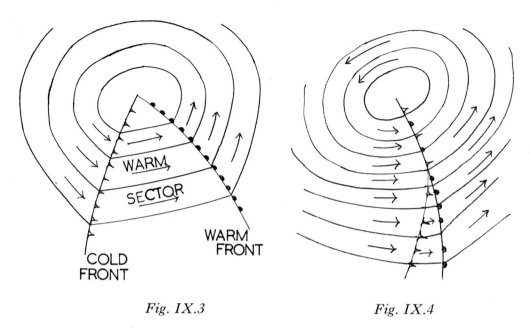

Fig. IX.3   *Fig. IX.4*

A typical section through the warm sector of a cyclone is shown in Fig. IX.5. The type of cloud to be expected is indicated. In the cold air before the arrival of the warm sector, the clouds are the normal fine weather cumulus. The warm front first arrives at any position on the earth high up in the atmosphere. The warmer air aloft produces an inversion which the columns of cumulus cloud cannot penetrate. As the depression approaches, the amount of vertical development in the cumulus is steadily diminished, and finally the cumulus clouds may be suppressed altogether. The first signs of the approach of a depression are usually to be seen in the advance of cirrus cloud – either the fibrous variety, or cirro–stratus, giving haloes round the sun and moon. The cirrus can first become visible when the warm front of the depression is over 1,000 kilometres away and it just reaches overhead when it is about 800 kilometres (say 500 miles) distant. The first signs of a depression can be detected in the clouds, often before the barometer shows any signs of falling. Similar signs may appear and then disappear and nothing may come of them, but if it is known from a weather forecast that a depression is approaching, then the appearance

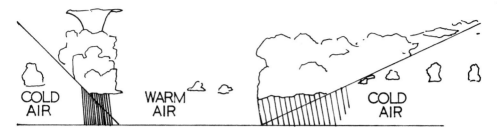

Fig. IX.5

of the cirro-stratus can give some ten or twelve hours' warning of the
time of arrival of the warm front and somewhat less, of course, of the
arrival of the rain.

The clouds associated with the warm front of a depression are caused
by the ascent by the warm air of the slope of the warm front, lying over
the cold air in front of it. The warm front is inclined upwards at a slope
of about 1 in 100 to 1 in 300. As the warm front approaches the clouds
thicken from high levels downwards. Finally rain reaches the surface of
the earth. It may cease as the warm front passes or continue in the humid
conditions of the warm sector. There is a veer in the wind as the warm
front passes (i.e. the direction of the wind swings round in a clockwise
direction).

The cold front which follows the warm sector is roughly some two to
four times as steep as the warm front. The cold air undercuts the warm
air in front and lifts it upwards. Since the same amount of lift is imparted
to the damp warm air in a shorter space, as was given to it by the warm
front over a distance three or four times as long, the effects produced by
the lifting are much more vigorous. They take the form of a line squall,
that is to say of a line of vigorous cumulo-nimbus clouds, often thundery
and with anvils. The rain is heavy but less prolonged and it is sometimes
called the clearing-up shower, because as the cold front passes by,
normal fine weather conditions finally return.

Except that there is no warm front at the surface of the earth, the
sequence of clouds attached to an occluded front is the same as that for
a warm front followed immediately by a cold one. Rain may or may not
reach the ground under an occluded front, depending upon the height
of the occlusion and the vigour of the depression.

In an anticyclone the air is slowly descending and being warmed as it does so. When it approaches the ground it spreads outwards. The warm air aloft makes for great stability. Cumulus clouds may not form at all, and if they do, there is little vertical development. The thin cumulus clouds of Colour Plate 12 are typical of an anticyclone in summer. The conditions are then settled and the fine weather is likely to last. Anticyclones unlike the cyclones, often anchor themselves in one position and move much less quickly. In Great Britain they tend commonly to last for about a fortnight once they have become established in one place, but elsewhere they may be a more or less permanent feature of a whole season. The subsiding air does not actually reach the surface because of convection and turbulence. Hence an inversion is likely to form which, in winter, produces stratus cloud and fog.

As cyclones follow each other across the country the regions of low pressure are necessarily separated by areas of high pressure. The weather in these high pressure areas is anticyclonic in type but less marked. It is normally fine but the high pressure does not remain stationary over one place, as it often does in an anticyclone. The fine weather is thus of shorter duration.

Observation of the clouds can be a useful adjunct to weather forecasting, even when a considered forecast is available. By keeping an eye on the clouds, it is possible to regulate activity out of doors so as to avoid the worst and make full use of the best that is available. The most difficult part of forecasting the weather lies in the timing of changes. Of this the clouds can give very useful indications. Judging the clouds can also provide valuable forecasts for short periods ahead. How often does one see, for example, picnic parties setting joyfully forth almost underneath the anvil of an approaching thunderstorm! A glance at the sky could easily save the family from a drenching.

The cyclones with which we have been concerned so far have been those of the temperate zones. Cyclones also occur in the tropics and there can take on a much more violent form. Why this should be so is not understood. The circulating winds have been pulled in towards the very low pressure in the centre, so that their speed is enormously increased. The circulating air is filled with towering thunderclouds. The most remarkable feature of the tropical cyclone, however, is the well-known calm 'eye' of the storm. Here the sky is clear and the sun may shine. At high levels the air is warmer by six or seven degrees as compared with the surroundings. This suggests that it might be possible

22. *Cloud streets. Warm spots on the ground in the distance lead to a series of bubbles of rising air, which produce cumulus clouds. These are carried along by the wind and form long lines or streets.*

23. *Cloud streets over southern England. The photograph was taken over London Airport from a height of 9,500 metres (about 28,000 ft) while in a queue waiting to land. The dark patch near to the reflection of the sun is the Isle of Wight, and the distant clouds on the horizon lie over the coast of France. The cloud streets extend from the south coast of England as far as the London area. The white marks in the upper part of the picture are reflections in the window of the aeroplane.*

24. *Cumulus clouds prevented from rising by an inversion. The inversion is marked by the layer of alto-cumulus cloud at the top of the picture. Below it are a number of cumulus clouds. Being insufficiently vigorous to penetrate the inversion they bend over underneath it.*

25. *Alto-cumulus lenticularis.*
This cloud is associated with a wave
in the interface between two different
layers of air. Such waves are often
formed when the air passes over some
obstacle such as a cliff or mountain
range. They are similar to the stern
waves generated by a boulder in a
stream. They remain stationary,
maintaining station with the obstacle.

26. *A lycopodium halo. The*
photograph is of an illuminated pin-
hole taken through a screen of lyco-
podium dust. There is a bright aureole
surrounding the pinhole. The pinhole
itself is obscured because it is impos-
sible to expose the film properly for
both it and the halo. Surrounding the
aureole is a series of coloured rings.
Many more can be seen by eye than
can be photographed, because of the
variation in brightness from one ring
to another.

27. *A mock sun. Mock suns are*
found at the same altitude as the sun
and near to the halo of 22°. They are
caused by refraction in ice-crystal
prisms sinking through the air with
their axes vertical.

that the eye of the storm could be accounted for by the rotation of the cyclone having reached such a high speed that the pressure in the centre has become low enough for air to be sucked into it from above. In descending it would be compressed and warmed. Other violent tropical storms which do not possess the warm eye may not develop into full tropical cyclones. Whether the eye is a result of the development of the cyclone beyond a certain stage, or is one of the causes contributing to the development is not, at the moment, clear.

The most violent of the aerial eddies is the tornado. A swirling column of cloud and debris sucked up from the surface of the sea or land, extends from the base of a heavy thunderstorm. Winds in tornadoes have been estimated to reach as much as 800 kilometres per hour (500 miles per hour) and pressure drops of as much as a tenth of an atmosphere have been recorded. Little seems to be known about how they come to be formed. They are often associated with violent line storms ranged along a line of rapidly increasing pressure, showing almost continuous lightning. Clearly, in very unstable conditions something triggers off a violent up-current in very humid, warm air, which may then be maintained by the liberation of latent heat of condensation. The air flowing in to take the place of the rising air spirals round in a vigorous swirl. Tornadoes occur comparatively frequently in the middle states of the USA and, although very limited both in duration and in area, they are very destructive of life and property which lie in their path.

Dust devils are miniature tornadoes but they lack the driving energy derived from the condensation of water vapour, which keeps the tornado going and makes it so violent. Instead of occurring in the stormy, humid and unsettled conditions in which tornadoes develop, dust devils are observed in quite the opposite kind of weather. They require very still, clear, hot days for their development. What is required is a flat, horizontal, dusty surface which becomes very hot indeed in the sun. The layer of air in contact with it also becomes very hot. The hot air lifts off the surface at certain points in convection currents, and the air flowing in to take its place spirals round. It is, in effect, the emptying of a 'bath' of hot air upwards, and the eddy forms near the 'waste pipe', carrying dust and leaves up in it. In Great Britain dust devils rarely develop to any size, though small ones are not uncommon in summer. The writer has seen some reach a height of over three metres in the Fens. In desert areas where the heating is much more intense, they are said to reach a height of 30 metres or more.

Dust storms, again, are different. Instead of developing in still conditions, they occur only when strong winds are blowing, which can carry the particles aloft. Sand storms, which are composed of comparatively large grains of sand, tend to be relatively shallow, the coarse particles not being lifted very far into the air. Dust storms, on the other hand, can rise to great heights and be carried for long distances. As has already been mentioned in Chapter II, dust from the Sahara was carried over Britain in the summer of 1968, and some of it was deposited here. It gave rise to a peculiar coloration of the sun. Instead of the familiar yellowish colour of the sun seen through high stratus cloud, the sun appeared much whiter. The soil of Great Britain is usually too damp and too thickly covered with vegetation to be carried away by the wind. Around the coasts sand can be carried by the wind for short distances, and it often gets built up into sand dunes. Some very light soils, such as those of north-east Nottinghamshire, can be blown off the fields during cultivation. In some areas of the world, however, wind erosion of the soil has been very severe, especially when unsuitable methods of cultivation have denuded the soil of the protection of plant cover, at a time when the soil was very dry. The so-called dust bowl of America is a celebrated case in point. The problem has been extensively studied there and special methods of conservation of the soil have been evolved.

Chapter X

Thunderstorms

The thunderstorm is the grand apotheosis of the atmospheric symphony, with all the instruments rendering the passage fortissimo. The thunderous percussion with its continuous rumblings and earth-shaking crescendos has probably been the most awe-inspiring of the many features of the thunderstorm, from time immemorial, and it is small wonder that it came to be attributed to the wrath of the gods. Thunder, however, is harmless, but not so the lightning from which it is derived. Thunder and lightning are the most striking characteristics of a thunderstorm. An elementary knowledge of electrostatics will furnish a good deal of the explanation, to consider which we will now turn. First, however, we must examine the state of affairs in calm weather away from any storm activity.

When a body is electrically charged the only manifestation of the fact is that other charged bodies experience forces when placed in its neigh-bourhood. It was Faraday who first concentrated attention on the space surrounding charged bodies. This space was called an electric field. There is an electric field in the atmosphere above the earth, even in fine weather. It corresponds to a negative charge on the ground and a positive charge on the ionosphere – the conducting layer which exists high in the atmosphere. The size of the fine weather field can be measured. If a conductor, such as a length of wire, is suspended in the atmosphere above the ground, it will gradually charge up to the voltage in the earth's field at that point, because the air is not a perfect insulator. The process can be hurried up by various devices. Lord Kelvin used a spray from the conductor (in the form of a vessel of water) to carry away induced charge and equalise the voltage of the conductor and that in the earth's field. (An electric field is measured in terms of volts per centimetre or metre. There is a voltage drop, or rise, from one point to another.) When equilibrium has been attained the voltage of the conductor can be measured by a suitable form of voltmeter. This is usually some form of electrometer, the design of which is described in books dealing with electricity. When measured in this way the earth's fine weather field is found to be surprisingly high – of the order of 130 volts per metre, near the ground. (The ordinary electric mains are at 240 volts in Great Britain.) At higher levels the field is smaller but it has been estimated, by measuring the field at various heights, that the ionosphere is about 360,000 volts positive relative to the ground. This high potential is across the atmosphere all over the world. The air is not a perfect insulator and its resistance can be measured. The resistance of a conductor is, of course, less the greater the area of cross-section through which the electric current has to pass. The resistance of the air lying between the ionosphere and the ground and covering the entire world, has been measured to be about 200 ohms. With 360,000 volts across this resistance it follows from Ohm's Law that we must have an electric current of 1,800 amperes flowing all the time. A current of 1,800 amperes at 360,000 volts delivers energy at the rate of 650,000 kilowatts – the output of a very large power station.

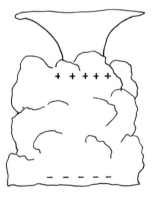

Fig. X.1

Underneath or very near to a thundercloud the fair weather field of the earth is reversed, the earth having a positive charge induced on it by the cloud. The ground being a conductor, its potential is, of course, the same everywhere. Under a thunderstorm the potential of the air is less than that of the ground, whereas in the open away from the storm, it is greater. There must be some mechanism for separating positive and negative electrical charges in a thundercloud, and the direction of the field under the cloud means that there must be a negative charge low down in the cloud, while the corresponding positive charge from which it has been separated must lie higher up.

If we know how far away a cloud is (for example by timing the interval between a flash of lightning and hearing the thunder, as will be mentioned later) and can make reasonable guesses about how far the positive and negative charges are apart in the cloud, it is not difficult to calculate the field to be expected from charges of any magnitude, Q. Each charge of Q units will produce a force on unit charge at a distance r from it of Q/r^2. The two forces (from the two charges in the cloud) will be slightly inclined and of slightly different sizes because of the difference in distance, and their resultant can be calculated – for example by the drawing of the parallelograms of forces. Having measured the electric field as before, we can calculate how big Q must be in order to produce it. It is in this way that the charges carried by thunderclouds have been measured. An average figure comes out to be something of the order of 30 coulombs. This order of charge is also neutralised when lightning connects the top and bottom of the cloud.

It is thought that thunderstorms form the generating units of the power station which supplies the normal earth's field and the 650 megawatts of leakage power through the atmosphere. It is thought that thunderstorms cover about one per cent of the earth's surface and that between 2,000 and 6,000 are going on at any one time. They produce some 100 to 200 lightning strokes every second. If these views are correct these thunderstorms will have to supply the 1,800 amperes of charging current to the ionosphere. Several times this amount must be lavished on the firework display of the lightning, since cloud to cloud strokes contribute nothing to the charging current. The 1,800 amperes are carried by lightning strokes to earth from the clouds. Now 200 lightning strokes per second each carrying 30 coulombs constitute a current of 6,000 amperes. Of these 1,800 amperes go towards the charging current of the ionosphere while the rest is devoted to the fireworks of cloud to cloud discharge. There must therefore be several times as many strokes taking place from cloud to cloud as from cloud to earth.

How the charges become separated in a thunderstorm is another matter. What we have discussed so far have been, in the main, facts determined by actual measurement. Many theories to account for how the thunderstorm works have been put forward. None of them seems to be completely satisfactory. One of the most plausible was put forward by C. T. R. Wilson, the inventor of the cloud chamber, in 1929. He supposed that in the cloud there are water drops of various sizes. The larger ones will fall through the air relatively faster than the smaller drops. Once a small charge has been established in the cloud, a drop will have positive charges induced on the lower surface and negative charges on the upper. As the drop falls it will present the positive charge to the air through which it is falling, and it will thus tend to collect more negative ions from the air than positive ones. The smaller drops will collect the positive ions left by the larger ones and take them to the top of the cloud. Some calculations have thrown doubt on whether Wilson's mechanism would transport the charge sufficiently rapidly, but the question remains unsettled. Another theory is that small drops lying in the path of a large drop, though not actually picked up by it but swept aside, might take the positive charge off the bottom surface of the large drop, just as the collecting combs of a Wimshurst machine collect the charge off the sectors as they pass close to them.[1] It has been noticed that the positive charge in a thunderstorm lies in the region of the cloud formed of ice particles, while the negative charge lies in the water drop region. For this reason another theory ascribes the charging of the cloud to charges generated when water freezes. It is known that ice, when it is formed, becomes charged relative to the remaining water. If the water were to be swept off the ice by the up-draught in which the ice particle was situated, a separation of charge could result. Further than this one cannot go at the moment.

A good deal of study has been devoted to the actual lightning flash itself. It has been discovered that a number of discharges, all following the same track at very short intervals of time, make up the lightning

---

[1] I first heard this theory by word of mouth from C. T. R. Wilson in 1923. He then maintained that the smaller drops might actually touch and bounce off the bottom of the larger drops, without being picked up by them.

stroke. This discovery was made with a camera invented by C. V. Boys, in which the film was spun rapidly behind the lens. Multiple discharges thus photograph on to the film as multiple images, and the course of the whole lightning stroke can be followed. The discharge is started by a faint leader stroke, which often passes down the track in stages. When it does so it is known as a 'stepped leader'. Occasionally the leader does the whole journey non-stop, in which case it is called a 'dart leader'. The leader is then followed by the main stroke, and this in turn may be followed by a number of others, up to twenty or more. The whole discharge may take as long as half a second. The individual discharges follow each other too rapidly to be distinguished by the eye, although sometimes a distinct flickering is to be observed.

The thunder which follows the lightning is nothing more than the sound of the discharge. The individual discharges take so little time to complete that the current in each is very large – of the order of 100,000 amperes. It is confined to a path which is estimated to be initially only a centimetre or less in diameter. Very high temperatures are produced – calculation indicates temperatures of 10,000° C or higher. The air along the discharge path expands explosively, and it is the sound of the explosion which is heard as thunder. In addition the succession of discharges, which follow each other to form the whole lightning flash, may give rise to sound of audio frequency.

The rumble of thunder may be caused in two ways. A flash of lightning may be about 3 kilometres long and it is not straight. Sound travels at 331 metres per second (about 1,100 ft per second) so that the thunder from a flash of lightning may take nine seconds to complete. Indeed, since the sound may be reflected from hills and other objects in the neighbourhood, the time may be even longer. Reflection in this way is the second factor which contributes to the rumble. Any part of the track which is more or less at right angles to the line drawn to the observer will send all its sound to his ear at about the same time. The sound coming from other parts will be spread out over an interval. The general result is the well-known variation in intensity which makes the rumbling sound.

The distance of a flash of lightning from the observer can be estimated by measuring the time between seeing the flash and hearing the resulting thunder. The light travels so fast that we see the flash almost immediately it takes place. The sound, on the other hand, takes three seconds to travel a kilometre (five seconds a mile) and the time between the flash and the sound in seconds divided by three gives the distance in kilometres or, if divided by five, the distance in miles. Thunder is rarely heard from lightning more than 5 or 6 kilometres (three or four miles) away.

The so-called summer lightning, sometimes to be seen lighting up the distant night sky in summer, but producing no thunder, is simply the light from the flashes of a distant storm, reflected by the clouds. (It is silent only because it is a long way off.) The same applies to 'sheet' lightning, except that it may be close enough for its thunder to be heard. 'Forked lightning' is the normal form of lightning discharge. It is seen when the actual path of the discharge is visible. 'Ball lightning' has been sufficiently frequently reported for its existence to be assumed, but it is nevertheless, very rare. It is said to take the form of a bright spherical discharge which travels comparatively slowly through the atmosphere, ending either by exploding or by simply fizzling out. It does not appear to be responsible for much damage but the rarity of its happening may perhaps account for this.

So much for the percussion instruments and their sounds: what of the percussion players – the air currents, and water as vapour, liquid and solid? The cumulo-nimbus cloud with anvil top, which we have already considered, is the initial stage in the formation of a thunderstorm. It is very often heard described in the weather forecast under the name of a 'shower of a thundery type'. There are some further features of the more mature storm with which it is worth the while of the person interested in the natural history of the clouds, to make an acquaintance.

The thunderstorm was investigated on a large scale in America by a team led by H. R. Byers. It consisted not only of ground staff at a network of stations, spaced out over a wide area, but also of air crews who flew their aeroplanes into and through thunderstorms at a series of different heights. The results were very extensive and were printed in a government publication entitled 'The Thunderstorm' (US Government Printing Office, 1949). Mr Byers also contributed an article to the 'Compendium of Meteorology' published by the American Meteorological Society of Boston, Mass.

What appears to happen in a thunderstorm may be briefly described as follows. The cumulo-nimbus cloud is, as we have seen, the seat of a vigorous up-current of air. The water droplets which it contains freeze at a certain level. Their latent heat is transferred to the air which supports them and relative to which they are falling. This latent heat is carried away by the air to higher levels, and gives rise to the anvil, as we have already discussed. The frozen drops grow by capturing supercooled droplets coming up with the further up-draught and thus a large mass of hailstones collects near the base of the anvil. As they become more numerous, the hailstones begin to block the passage of the up-draught and slow it down. At length so much ice is accumulated that it can no longer be supported, and it falls out of the cloud, carrying a good deal of air with it. The term 'cloud burst' is not altogether too exaggerated to describe what actually happens. At some point the ice may melt, absorb its latent heat and cool further the air which is in contact with it, thus increasing the down-draught. Finally the cold air spreads out in a layer over the ground, which may be 600 to 1,500 metres thick (2,000 to 5,000 ft).

The cold air possesses the velocity of the wind at a high level – the so-called 'directing level' of the storm – and spreads out on the ground mostly in this direction. It gives rise to a miniature cold front, forcing its way underneath the warmer air before it, and causing the warmer air to rise. In the fairly unstable conditions in which thunderstorms arise, this produces further vigorous up-currents, in the same manner as the

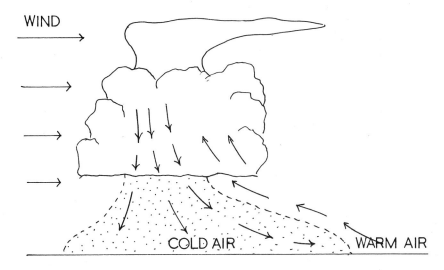

WIND

COLD AIR   WARM AIR

Fig. X.3

cold front of a cyclone, and tends to maintain the storm. Storms thus may become self-propagating in this way. The structure of a mature thunderstorm is as sketched in Fig. X.3. After the mature stage has reached its climax there follows the stage of dissipation. The down-draught spreads throughout the area and the up-currents cease. Layers of stratus cloud derived from the anvil may remain and shut out the sun. The aftermath of the storm is thus a sudden change in conditions from hot sultry weather to cool and cloudy.

The pattern of the rainfall follows the storm cell. The most intense rain occurs underneath the centre of the cell and usually takes place within two or three minutes of the onset of the rain. Heavy rain lasts for from five to fifteen minutes as the cell passes over, after which the rain slowly gives way.

The winds associated with a thunderstorm are such as to account for the common experiences of a storm 'advancing against the wind' and of 'the calm before the storm'. The storm, being the seat of up-currents which furnish it with its supply of energy, must draw in air from the surroundings to replace that which has gone up into the storm cell. Just outside the storm area, therefore, the winds will be directed inwards. These will be on top of the general wind already existing. Thus, in front of the storm the in-draught will be in the opposite direction to that of the air in which the storm is drifting. The two may neutralise each other and produce the 'calm before the storm'. Nearer the storm the in-draught may appear the stronger so that the storm itself will advance, apparently against the surface wind which is blowing towards it. There is a change in the direction of the wind as the storm gets near – as soon as the storm winds arising from the outflow of the cold air of the down-draught arrive. These winds can be very strong.

A thunderstorm is often said to 'clear the air'. This is the effect of the pool of cool air left behind and produced by the down-draught. The temperature may be reduced in this way by as much as 10° C to 15° C. The area affected by the cooling is greater than that over which intense rain occurs, because the dome of cold air underneath the storm spreads out, and it continues to do so after the storm itself has ceased.

There is one rather curious feature of the weather which is often associated with a thunderstorm. The humidity, which is usually high in the weather preceding the storm, actually drops sharply at the time of the heaviest rain. Condensation in hailstones or cold raindrops may desiccate the air, or the effect may be due simply to the rate of evaporation of the raindrops in the descending air not being sufficient to keep the air saturated.

The very vigorous up-currents in a storm are able, by their buoyancy and momentum, to penetrate for a considerable height into the stratosphere. The American thunderstorm project found some ascending as high as 20,000 metres (60,000 ft). However, high-level flight remains the best way of avoiding thunderstorms, while the worst choice of flight path remains within the storm.

Another very notable feature of thunderstorm rain is the rate at which it falls. The raindrops and hailstones can fall much more rapidly in the descending air, which itself is going down at a considerable speed, than they would in still air.

Chapter XI

Haloes and Rainbows

There are two quite different kinds of halo which form round the sun and the moon. We will consider first of all the kind which forms very close to the discs of these objects. These are to be seen most easily around the moon, the sun being too bright to be viewed comfortably, except through smoked glass. When either the sun or the moon shines through a layer of thin high cloud, the area just outside the disc appears bright; in the case of the sun, it is very bright, like burnished brass. This central area is known as the aureole. Its outer edge is usually of a reddish tinge. Surrounding the central area a series of rings may sometimes be seen. These rings are known as the corona and they are bluish on the inside and reddish on the outside. The aureole is the part of this type of halo which is most often to be seen. It is not at all uncommon. The rings are seen less frequently and it is unusual for more than one to be visible. As many as five have, however, been recorded.

This type of halo is caused by the process which is known as the diffraction of light. It occurs whenever light has to pass near an obstacle. In the case of the halo it is the passage of light from the sun or moon past the water droplets of the clouds.

Light possesses the properties of a wave motion. If two equal trains of waves arrive at a point so that the crests of one train arrive at the same time as the troughs of the other, the two trains will interfere and destroy each other. In Fig. XI.1 waves are represented passing a small obstacle. We would obtain the same effect if we removed the obstacle and generated waves through its surface, the crests of which fitted on to the troughs of the original waves and vice versa. We see from this, therefore, a result which is very surprising at first sight, namely that in directions in which the original light produces no illumination, the effect of a small obstacle is just the same as that of light passing through an opening in an opaque screen, the opening having the same outline as the obstacle. This fact is known as the Principle of Babinet, after its discoverer.

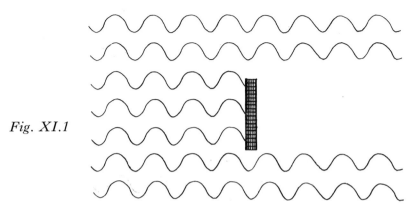

Fig. XI.1

Let us look, therefore, at the illumination to be expected when plane waves pass through a rectangular opening. Such waves produce illumination not only in the direction in which they are travelling but, because of diffraction, in other directions close by, as well. As we move round from the original direction of propagation OX (Fig. XI.2) we shall come to a position, in the direction OY, where the disturbance from one half of the opening OB, arrives at a distant point, having travelled half a wavelength further than that from the other half OA. From these two halves the crests from one will arrive at the same time as the troughs from the other, and the two waves, which would obviously otherwise be equal to each other, will destroy each other. Darkness will result. Further round still than OY we shall come to a position where the path difference will be a whole wavelength. Crest will fall on crest and brightness will result.

The theory for a small circular opening is very nearly the same, and if a screen is placed to the right, on which the light can be received, we shall see a bright central area surrounded by a series of bright and dark rings. The same pattern is to be expected, according to Babinet's Principle, from a small obstacle. The angle of the bright rings obviously depends upon the size of the object giving rise to the diffraction. With a small obstacle we have to go further round before there is half a wavelength difference between the two half waves. We see from Fig. XI.2 that when there is a half wavelength difference between the two half waves, the length BD must be equal to one wavelength, λ. We see from Fig. XI.3, therefore, that the angle of diffraction is greater for a small

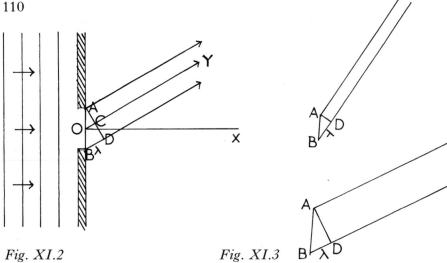

Fig. XI.2   *Fig. XI.3*

obstacle than for a larger one. In fact, because the wavelength of light is very small, unless the obstacle is also very small, the angles of diffraction are too small to be noticed. The wavelength of light is about five hundred-thousandths of a centimetre. The angle of diffraction to the first dark ring will be one in ten when the diameter of the obstacle causing the diffraction is about five ten-thousandths of a centimetre. The droplets to be found in clouds run from about this size to those some twenty times in diameter. The angles to be expected in the haloes are therefore likely to be of the order of a few degrees.

If a large number of small obstacles, all of the same size, are scattered in the path of a beam of light, each particle will form maxima and minima of brightness in the same directions. This is thought to be the origin of the aureole and corona. The reason why many rings are seldom seen is that it is rare for the particles in a cloud to be all exactly of the same size.

The most beautiful haloes can be produced in the laboratory very easily. Most commonly occurring fine powders, such as French chalk, consist of irregular fragments of a variety of sizes. They produce only an aureole when introduced into a beam of light. Lycopodium powder, however, is the exception. It is formed of the spores of the club moss lycopodium, and was formerly used by chemists for the coating of pills, but chemists' shops nowadays do not stock it. It can, however, be obtained from laboratory suppliers quite cheaply and most physics laboratories have a supply of it. The spores are more or less spherical in shape and very uniform in size. They produce magnificent haloes consisting of many rings of very brilliant colours.

To produce the halo put a small pinhole, pierced in a piece of metal foil (such as chocolates or cigarettes are wrapped in) in a slide projector and focus an image of it on the screen. Dust a glass plate lightly with lycopodium powder and place it in the path of the beam of light from the projector, close to the lens. The haloes will be seen surrounding the image on the screen. A translucent projection screen is better for viewing than an opaque white one. Directions for making a better diffracting screen than can be obtained by simply dusting the powder on a glass plate, will be given in the next chapter.

The second type of halo to which we have referred is formed not by diffraction but by refraction – refraction in ice crystals. The diffraction haloes can be formed in either water drop or ice crystal clouds. Refraction haloes can only occur in clouds formed of ice crystals. The common halo of 22° radius belongs to this class.

Refraction is the bending of light as it passes from one medium to another, and the passage of light through a prism is a well-known example of it. The 22° halo is formed when light from the sun is refracted in an ice prism, the angle of which is 60°. Ice crystallises in the hexagonal system. The angles between the faces of a hexagon are 120°. Light cannot

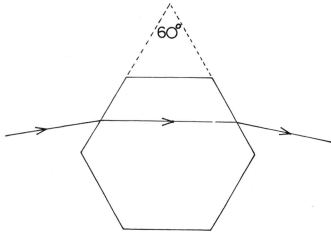

Fig. XI.4

pass through a prism of ice of this angle. All the light which enters by one face will be reflected by the second and none will pass through. However, if light enters by one face it can leave by the next but one. If it does so, in effect it passes through a prism the angle of which is 60°, as can be seen from Fig. XI.4.

We can reproduce what happens in the laboratory very nearly, if we use a prism of water instead of ice. The refractive index of ice is 1·31, whereas that of water is 1·33, very nearly the same. We have to use a hollow 60° prism to hold the water, of course. This can be purchased or one can be made by cementing perspex sheets together, or by sticking glass plates together with sealing wax (Plate XI.1). The latter will not stand being dropped on the floor but the prism is otherwise surprisingly robust if properly made. Details of the method of making it will be given in the next chapter.

Plate XI.1   A 60° prism made by cementing lantern plates together with sealing wax.

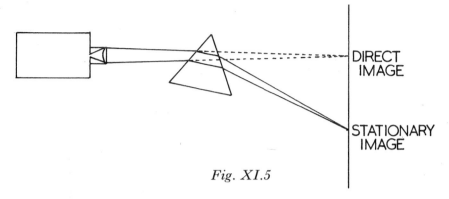

Fig. XI.5

To experiment with the prism it should be stuck firmly to a piece of wood which can be mounted on a small electric motor and spun round. Care must be taken to see that the prism spins about its mid-point with the sides vertical and, of course, it must be provided with a lid or the water will spin out. The prism can be mounted on a gramophone turntable although the speed is hardly sufficient to show the effect clearly.

Place a narrow slit in the slid projector and focus its image on a screen. Place the prism, mounted on the motor, so that the beam of light passes through one of the angles of the prism. When the prism is rotated the refracted image first moves towards the position in which the direct image was focussed in the absence of the prism, but it does not reach it. It comes to rest and then returns again and moves away. The same thing is repeated as the next angle comes into the path of the beam.

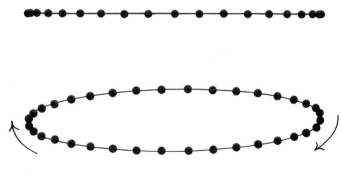

Fig. XI.6

The effect is very similar to what would be observed if a light were to be mounted on a rotating wheel viewed in the edge on position, as in Fig. XI.6. When the rays to the eye are nearly tangential to the wheel, the light appears to be almost stationary. The same thing can happen when an ember from a bonfire is whirled round on a string in the dark. The light appears to be concentrated at the points where it is momentarily stationary. The appearance is the same with the spinning prism. The position where the image is momentarily stationary is known as the position of minimum deviation, because the beam of light is deviated least from the direct line. We can measure the angle of minimum deviation by drawing lines on a piece of paper from the position of the centre of the prism to the stationary position of the image and also to the position of the undeflected image. The stationary position is slightly different for lights of different colours, the red lying towards the direct image. If we measure the angle from the direct image to the red stationary image we shall find it to be about $23\frac{1}{2}°$. Much will depend upon how accurately we have made our prism. However, using lantern plates joined by sealing wax, it is not difficult to make a good 60° prism.

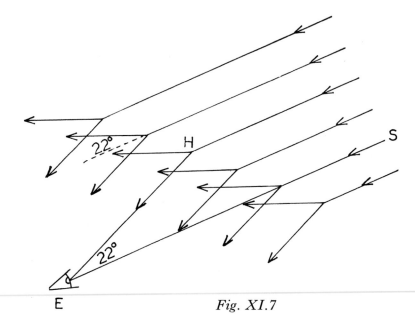

Fig. XI.7

As ice crystals tumble through the air they are likely to be arranged in all possible directions. The effect of this, and with a very large number of prisms, will be very similar to what we have obtained with one prism by spinning. The light will be concentrated in the direction of minimum deviation making an angle of D say, (which for ice is 22°) with the direct beam.

When a cloud of a very large number of ice crystals is illuminated by sunlight, each small volume of the cloud will send out a cone of light which has been refracted by the prisms (Fig. XI.7). The half angle of the cone will be 22°, and it will be centred on the line joining the apex to the sun, as axis. An eye, E, looking towards the sun, will also receive the refracted rays from a ring round the sun, HH, lying in a direction 22° away from the sun. In this mechanism, therefore, we have found the way the common ice halo is formed. Since some light is deviated through an angle greater than 22° (but none less than this) the region outside the halo will appear to be brighter than that lying inside. This can be seen very clearly in Colour Plate 2 and furnishes an additional confirmation of the theory.

Since the angle of minimum deviation is less for red light than it is for blue, the halo is coloured red on the side towards the sun. The angle of minimum deviation is different for different colours and we might expect that the halo would show the colours of the rainbow. The colours overlap each other, however, and they are very impure. Only the red is deviated on its own and the other colours are very pale and much mixed with white.

A common shape for an ice crystal is a flat hexagonal plate as in Fig. XI.8. When such a plate falls through the air, if of sufficient size, it tends

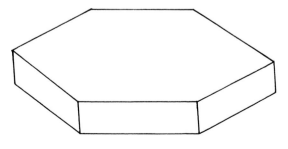

Fig. XI.8

to do so with its plane horizontal. Such crystals falling in this attitude will deviate the light horizontally.

If there are a number of such plates in a cloud, the halo tends to be reinforced in positions at the same altitude as the sun. These patches of stronger light are known as 'mock suns'. They are not at all uncommon. A photograph of one is reproduced in Colour Plate 27. Mock suns are best seen when the sun is low in the sky. If this is not the case the mock sun lies a little outside the halo because the rays of light do not pass through the ice prism in the plane at right angles to the refracting edge. The effect of an inclination in the rays of light is to increase the angle of minimum deviation, as can be verified with the help of the water prism. Mock suns cannot be formed if the sun is higher in the sky than about 60° (actually the limit has been calculated to be 60° 45′). At this altitude the sun cannot send any light through the faces of a horizontal crystal plate which make an angle of 60°. All is reflected and none transmitted.

When the sun is about to set, or has just set, a vertical pillar is sometimes seen to arise from it. This is caused by reflection in the horizontal faces of flat ice crystals, just as there is a brilliant glittering streak of reflected light from the sun in rippled water. If the ice crystals were all strictly horizontal as they fell there would be an image of the sun to be seen, as in a mirror. Very strong reflections, approaching this state of affairs, have been seen from aeroplanes flying above ice crystal clouds. However, the alignment of the crystal plates is usually not perfect and the effect is more nearly that of the glitter path on water. The individual ripples are represented by the separate ice crystals and they are too small to be seen separately. The pillar vertically above the sun is caused by reflection from the under surfaces of the crystals, while pillars below the sun arise from reflections in the upper surfaces. The reflections can be internal or external.

The commonest of the halo phenomena to be seen are those described above, but they do not exhaust by any means the range of those which have been seen from time to time. The remainder of those which we have not discussed are too complicated to go into in this book. They are all comparatively rare phenomena and the theory of their causes is by no means completely settled. What holds up the investigation is not so much the complication of the theoretical study as the paucity of reliable observations. Here the layman could play an important role. If he knows what to look for and can photograph any case which he is fortunate enough to see, noting the focal length of his lens and the elevation of the sun, so that angles can be estimated on his film, he would be able to make a highly valuable contribution to the study. A full development of all the ice particle haloes is extremely rare, but parts are occasionally to be seen. In order that the reader may know what to look for we will describe briefly an exceptionally full development which was recorded by a Russian astronomer, Tobias Lowitz, on 18th July 1794 in St. Petersburg. This classical occurrence, which contains a large proportion of what can be seen, is known as the Petersburg phenomenon. Lowitz commenced observation at 7.30 in the morning, and the fullest development occurred two and a half hours later, at about 10 o'clock.

Lowitz's drawing of the entire heavens, showing all the arcs to be seen on that occasion, is the basis of the sketch reproduced in Fig. XI.9. The 22° halo is the small circle surrounding the sun. It is surrounded by another, which appears to be elliptical in shape, known as the circumscribed halo. The two mock suns lie just outside the circumscribed halo and are joined to the 22° halo by arcs known as Lowitz arcs, because he was the first to record them. A horizontal white circle passes through the sun and the mock suns. It is known as the mock sun ring or parhelic circle (mock suns are also known as parhelia, and sometimes as sun dogs). Outside the 22° halo and its associated arcs and parhelia lies another halo, complete except for the interception of the horizon. Its radius is about 46°. It corresponds to the angle of minimum deviation in an ice prism of angle 90° – the angle between the base face and the side faces of the hexagonal prism. Touching both the 22° halo and that of 46°, are to be seen tangential arcs known as 'arcs of contact'. Finally, there are two arcs passing through the sun and known as 'oblique heliacal arcs'. The 22° halo and its mock suns are common enough, but photographs of any of the other features would be of great value.

The point on the parhelic circle opposite to the sun is known as the anthelion. It is, of course, at the same elevation above the horizon as the sun itself. Oblique arcs passing through the anthelion have been occasionally observed. They can be accounted for by reflection from the faces of long hexagonal ice prisms falling with their long axis horizontal. Now that travel by aeroplane is a commonplace, arcs below the horizon are occasionally reported. Reflection in the horizontal faces of ice platelets will give a mirror image of some of the phenomena occurring above the

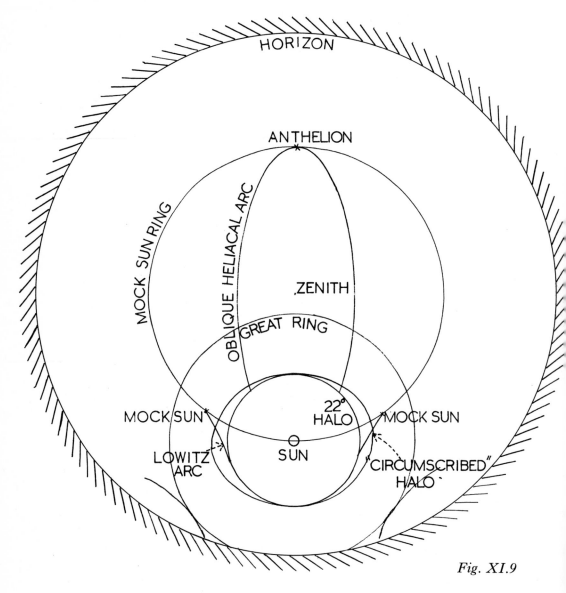

Fig. XI.9

horizon. A horizontal arc passing through the point immediately below the sun, and the same distance below the horizon as the sun is above it, has been reported. It is a reflection of the mock sun ring. Mock suns lying on this ring have also been seen. The point below the sun through which the arc passes is known as the sub-sun. The point opposite to the sun on the celestial sphere is known as the anti-solar point. It must not be confused with the anthelion. The anti-solar point is the point around which one's shadow will be centred. When flying, the shadow of the aeroplane will be at this point. Arcs passing through this point can also occur but its main interest is that it can be the centre of a rainbow-like system of rings known as the 'glory' to which we shall refer later. The glory is formed in water drop rather than ice particle clouds. The anti-solar point lies below the horizon so long as the sun is above, in contrast to the anthelion, which is above the horizon.

Of all the optical phenomena connected with the clouds and the rain which falls from them, the rainbow is the best known and probably the most beautiful. Interest in it goes back to time immemorial, but it was not until Descartes (1596–1650) that the theory began to be understood. Even today the theory cannot be looked upon as entirely completed. Descartes, however, took the first and most important step forward and for a long time his theory seemed so complete that interest in the subject dropped, although much fuller and closer observation was really necessary. Descartes traced the paths of a large number of rays refracted into a raindrop, reflected off the back surface and then refracted out again as in Fig. XI.10. He found that the angle, D, through which the

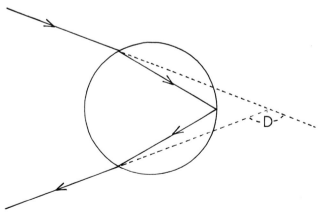

Fig. XI.10

ray was deviated, went through a certain minimum value, just as we found to be the case with refraction in an ice crystal. As in that case, the light which emerges from the raindrop is concentrated in the direction of minimum deviation. Using a refractive index for water of 1·33, the angle of minimum deviation comes out to be about 138°.

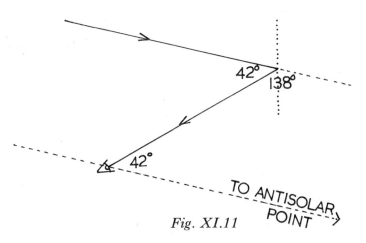

Fig. XI.11

We see a rainbow when the sun illuminates raindrops falling from a cloud. We must have our backs towards the sun and the circles of the rainbow appear to be centred on the anti-solar point. That is to say, the centre of the bow lies on the line from the sun which passes through the eye. We see from Fig. XI.11, that the rainbow will have, according to this theory, a radius of about 42° (the refractive index of 1·33 is appropriate for light of a yellow colour refracted in a water surface). Because angles of deviation greater than the minimum value of 138° occur, we see that the area of the sky lying within the rainbow should be brighter than that outside. This seems to be the opposite to what occurs in the case of the halo of 22°. The explanation of the difference is that the extra illumination lies in the area of the sky away from the sun in both cases. With the rainbow, however, the deviation is greater than 90°, and so the area away from the sun appears to lie within the rainbow.

The colours of the rainbow: red, orange, yellow, green, blue and violet (sometimes indigo is inserted between blue and violet) are well known. Red light occurs on the outside of the bow and the blue on the inside. Outside the primary rainbow a fainter bow is often to be seen.

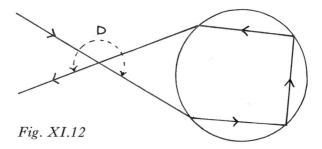

Fig. XI.12

It is known as the secondary bow. Descartes also found the theory of this bow. He explained it as caused by light which has suffered two reflections in a raindrop before emerging (Fig. XI.12). Here the angle of minimum deviation (marked D in the figure) comes out to be 231°. Light emerging after two reflections will thus be concentrated on a cone of semi-vertical angle 51° (321°–180°), as can be seen from Fig. XI.13. Fig. XI.14 then shows that the angle of the secondary bow is also 51°. The secondary bow will thus lie outside the primary.

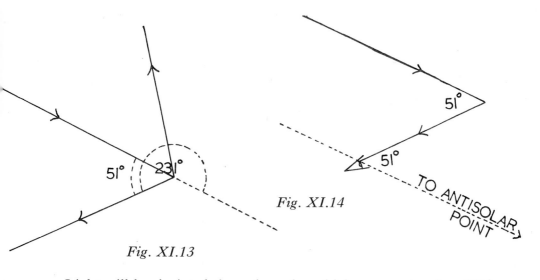

Fig. XI.14

Fig. XI.13

Light will be deviated through angles which are greater than 231°, which will correspond to an angular radius from the sun of greater than 51°. The sky thus appears brighter outside the secondary bow than inside it. There is therefore a space darker than usual between the two bows.

Minimum deviation in the secondary bow is less for red light than for blue, just as was the case in the primary bow. The angle of the primary bow is 180° minus the angle of minimum deviation and the angular radius of the red bow is greater than that of the blue. In the secondary bow the angular radius is the angle of minimum deviation minus 180°, and is therefore less for red light than for blue. The two bows lie with their red edges facing each other.

Faint bows are also often to be seen lying inside the primary. They are known as 'supernumerary bows' because their existence was not accounted for by the Cartesian theory. They are caused by diffraction. When light passes a straight edge the shadow is not sharp, because of the wave properties of light. Some light bends into the geometrical shadow, OG (Fig. XI.15) and there is a gradual increase in illumination towards

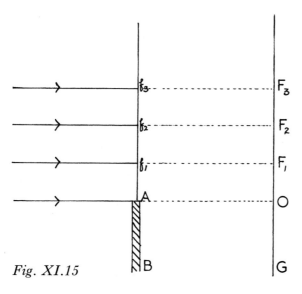

Fig. XI.15

O. More remarkable, however, is the fact that in the area which would be illuminated according to geometrical optics, there occur a number of bright and dark fringes. These arise because the point F_3, for example, is illuminated by a complete half-wave passing the obstacle, extending from f_3 upwards, together with the portion f_3A of the other half-wave. In the absence of the obstacle the screen would be illuminated, of course, by both halves of the wave front. The parts of the wave front, such as f$_3$A, would, on their own, produce disturbances which are not in step with those from the upper half-waves. To produce the diffraction fringes of the straight edge the disturbances from the portions such as f$_3$A, are combined with that from the other half of the wave. The result is, as is easily imaginable, still a series of fringes. A graph of the illumination of the screen is drawn in Fig. XI.16. The edge of the geometrical shadow corresponds to the point O. At this point the intensity of the illumination is only a quarter of that well away from the shadow.

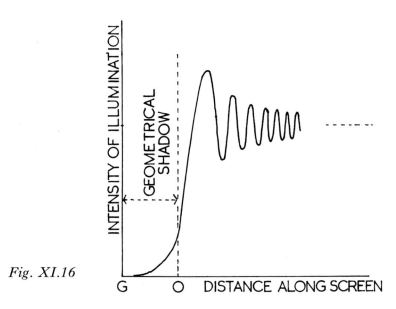

Fig. XI.16

The diffraction from a raindrop arises in a similar fashion. The effect of the raindrop is to produce a reflected beam in the form of a cone of light which is sharply cut off, according to geometrical optics, on the outside. In a manner similar to the shadow of the straight edge we find a series of bright and dark fringes in the area which would be illuminated according to geometrical optics, that is to say inside the cone which is blocked off. This region corresponds to a larger angle of minimum deviation and a smaller angular radius measured from the anti-solar point. The bright fringes lie within the main rainbow and give rise to the supernumerary bows. The colours of these supernumerary bows are

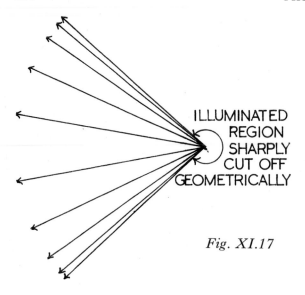

ILLUMINATED
REGION
SHARPLY
CUT OFF
GEOMETRICALLY

Fig. XI.17

usually faint pinks and blues, although there is some variation according to the size of the raindrops.

The last of the optical phenomena connected with clouds which we will discuss is that called the 'glory', to which we have already alluded. It is sometimes called the Brocken Spectre, after the peak in the Harz mountains from which it has frequently been seen. It was first remarked upon by Bougeur and La Condamine and a Spanish captain named Ulloa, in the Andes, when on an expedition to measure the length of a degree of longitude, in 1735. When one's shadow falls upon a cloud of water droplets, the shadow of one's head sometimes appears surrounded by rings of coloured light. Nowadays it is a very common experience when travelling by air, the shadow of the aeroplane on a cloud often showing the phenomenon. This is another result of diffraction of the light emerging from a raindrop after reflection from the back surface. The supernumerary bows are also formed by light which has been diffracted from the wave front near to the Descartes' ray (the Descartes' ray is the ray suffering minimum deviation and it is the origin of the rainbow according to geometrical optics). The supernumerary bows, however, are weak and soon become too weak to see a little distance from the primary rainbow. A very little light, however, will be diffracted backwards in the direction in which it came from the sun before entering the drop. In this direction, however, diffraction from the whole cone of emerging light will be of equal importance and not merely that arising from the wave front near to the particular Descartes' ray corresponding to the observer. There is thus a great increase in the illumination of the supernumerary bows near to the backwards direction – that is to say around the anti-solar point. It is this backwards diffracted light which gives rise to the glory.

The glory is to be seen surrounding the head of the observer only. If the shadows of his companions are also visible to him on the cloud they will show no glory. Just as everybody sees his own rainbow only and each person sees a slightly different bow to those seen by his companions, each person has his own glory. If the glory is photographed by a camera, it will be found to surround the shadow of the camera. There is no virtue attaching to the shadow of the observer's head except that it is the shadow of the place where his eyes are set! If the camera is held at waist level the glory will surround his stomach in the photograph, even though the observer will himself see it round his head, when he takes the picture.

The rings of the glory usually show colour with the red on the outside and the blue on the inside. This corresponds to the longer wavelength of the red light compared with that of the blue. Although the commonest appearance is of one or two rings only, as many as seven have been recorded. The centre is usually bright though the relative intensity compared with that of the first ring seems not to be constant.

Chapter XII

Simple Things to Make, to Think About, and to Do

In this concluding chapter we shall make a few suggestions for some things which the amateur can attempt to do. A study of clouds which requires no activity on the part of the student, like any other passive study concerned with a scientific topic, would be a poor thing compared with one in which more than an armchair study is involved. For convenience of reference the suggestions will be numbered but this is not intended, in any way, to indicate an order in which they should be undertaken. It is not essential that all, or indeed any of them, should be acted upon. It is a list from which a selection may be made if desired.

GENERAL CHARACTERISTICS OF CLOUDS

1. *Making a Collection of Cloud Photographs*   This is one of the most rewarding activities which can be indulged in when studying clouds, and it is an extremely easy thing to do. An elaborate camera is not required. The clouds are almost always brilliantly lit so that an expensive lens, with a wide aperture to let in a lot of light, is not required. Since the lens will almost certainly be stopped down, almost any lens will give good definition. If colour film is employed no special precautions, other than giving the correct exposure, are necessary to bring out the clouds in the picture. If black and white film is employed it is necessary to make sure that it is panchromatic and a deep yellow filter should be placed in front of the lens. The filter shows up the clouds by making the blue sky appear dark.

A very convenient filter can be made by sticking with sealing wax two cleaned glass lantern plates or cover glasses together, about a millimetre apart, to form a cell. It is not difficult to make a watertight joint and the result is surprisingly robust. The space between the plates is then filled with a solution of potassium bichromate, by means of a fine glass tube. The opening through which the cell was filled is then sealed off with a little more sealing wax, and the result is a clear filter which will not deteriorate, and which, with reasonable care in handling, is capable of lasting for years. [1] A series of filters, using solutions of different strengths, can be made, though one of fairly deep colour will suffice for most purposes. Alternatively, a filter can be bought from the photographic dealers. Details about how to make a sealing-wax joint will be given later (see No. 37, p. 136). All the photographs in Plate XII.1 were taken with a bichromate filter made in this way.

It takes about two years at least to make a fair collection of cloud photographs, and a camera must always be to hand throughout that time to take advantage of whatever offers. Many effects are very fleeting and must be taken as soon as they occur.

When taking photographs of haloes or of clouds towards the sun, it is well to arrange that the sun itself is screened off by a convenient tree or chimney pot. Even so an exposure about two stops smaller than that indicated by a light meter pointed in the direction of the photograph, will be sufficient. Otherwise the halo may show up very poorly or not at all.

2. *Hyperstereograms of Clouds*   Further interest can be given to cloud photography by taking stereoscopic pictures. Stereoscopic vision depends upon the slightly different views of the landscape which are obtained by each eye, because they are separated in space. It is effective over distances of less than a kilometre. Our eyes are not separated sufficiently, however, to enable the shape and spacing of clouds to be detected by this means. By photography it is possible to separate the viewpoints and produce exaggerated stereoscopic pictures, which can show up many points of interest that would not be obvious otherwise. In the case of the clouds, stereoscopic pairs of pictures can be obtained with a single camera. The clouds drift by on the wind, and if two pictures of the same clouds are taken across wind at an interval of two or three seconds (just about as long as it takes to wind the film on) there will be a relative displacement of viewpoint between the two, which may correspond to a separation between the eyes of ten or twelve metres or more. Some examples are reproduced in Plate XII.1. They have been printed at roughly the correct inter-ocular distance and can be seen stereoscopically without a stereoscope, given a little practice. Hold a piece of card or stiffish paper (black is preferable) at right angles to the paper along the line joining the photographic pair, so that the right eye does

---

[1] The use of boiled water for the solution will prevent the formation of air bubbles in the filter after it has been sealed.

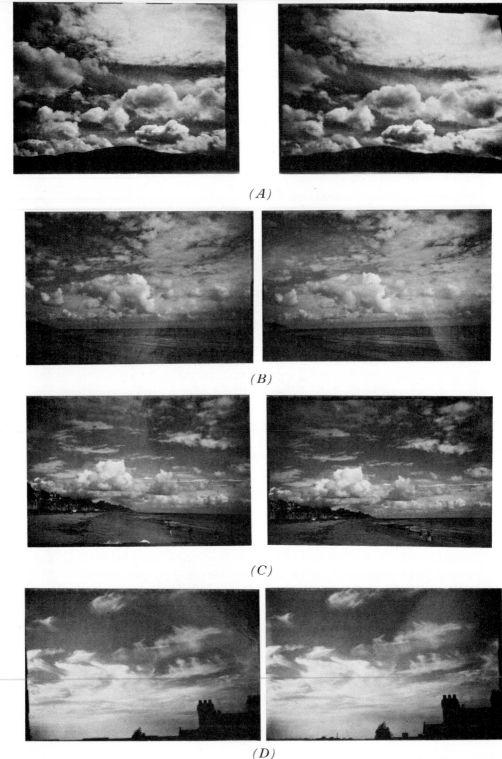

(A)

(B)

(C)

(D)

Plate XII.1

Plate XII.1   Hyperstereograms of clouds. Photographs of clouds taken across wind at an interval of two or three seconds so that there is a relative displacement of the clouds between each pair. Each pair of photographs forms an exaggerated stereoscopic combination corresponding to a distance between the eyes of twenty or thirty metres. The reproductions have been mounted so as to be at roughly the correct interocular distance and can be viewed stereoscopically without the aid of a stereoscope. Place a piece of black card vertically between the pictures so that each eye can see only one picture. Some difficulty may be experienced in focussing the eyes on the pictures at the same time as keeping the axes of the eyes parallel. A pair of convex lenses (about 10 cm focal length), taped over the holes in a postcard to act as spectacles, can be very helpful in surmounting this difficulty.

(A)   In the first pair three distinct layers of cloud will be visible – cirrus, alto-cumulus and cumulus.

(B)   This pair of photographs demonstrates the existence of a definite cloud base. All the cumulus clouds have their bases at the same level although they are of very different sizes.

(C)   This pair is similar. The landscape, which is the same in both members of the pair, appears at infinity.

(D)   Cirrus cloud. The interval between this pair of pictures was longer, corresponding to the greater height of these clouds. A curtain of falling ice crystals will be seen descending from the clouds on the right-hand side.

not see the left eye's picture and vice versa. The two pictures will suddenly 'click' into place and the three-dimensional view can be very striking. Seeing stereoscopic pictures without the aid of a stereoscope is easier for those who are short-sighted, as they are not worried by problems of focus. Those with normal sight tend to alter the focus of their eyes automatically according to the degree of convergence. They can get over the difficulty by looking through convex lenses, which can be taped on to a postcard in which two holes have been cut, one for each eye, about six to seven centimetres apart. If the lenses are of about 10 cm focal length, some magnification will also be obtained.

The pair of pictures in Plate XII.1 (A) shows three distinct layers of cloud. That in Plate XII.1 (B) illustrates very clearly how the base of cumulus clouds occurs at a certain level. The large one has its base at the same level as the smaller ones. The photographs of Plate XII.1 (C) show uniform heights of the cloud base. Plate XII.1 (D) shows cirrus cloud. Being further away, a longer interval of time between the two exposures is required. A curtain of precipitation can be seen falling from the cloud at about 2 o'clock from the centre of the picture.

3. While still on the general characteristics of cloud an excellent study is to examine the clouds portrayed by Turner or by Munnings. Both were superb students of nature and their paintings will repay careful examination to see what were the points which they noticed and tried to record. Reference to Ruskin's *Modern Painters* is also very interesting.

HUMIDITY

4. Seaweed hung up out of doors, but screened from the rain, absorbs moisture when the air is damp and becomes limp. When the air is dry it becomes hard. It can be used as a rough indicator of humidity – good enough to indicate when there is a good 'dry' for doing the laundry.

5. Dissolve some cobalt chloride in water and soak some filter paper in the solution and allow to dry. Cobalt chloride is a pale pink when damp and blue when dry. Its use as an invisible ink has been known for 250 years. Experiment to see whether cobalt chloride paper is superior to or of less value than seaweed as an indicator of the humidity. If a graduated hygrometer is available (these can be purchased inexpensively) it would be of interest to find out at what humidity the colour changes.

6. Make a 'weather châlet' by mounting the arm holding the man and the woman on a piece of catgut stretched by a small weight.

7. Make a hair hygrometer. A hair absorbs moisture and increases in length in damp air. To use it to make a hygrometer, the film of oil, which normally covers a hair, must be removed by boiling in a solution of washing soda for half an hour, after which the hair should not be handled or roughly treated. It should be fastened at one end and allowed to hang vertically, stretched by a small weight. Near the lower end it should be attached to a light lever, near to the pivot, so as to provide a magnification of any change in length which may occur.

8. Measure the dew point. This is a most important measurement. It can be done approximately by polishing a copper calorimeter (it is better if it can be silver-plated) and filling it about a third full of water. The water can then be slowly cooled by dissolving hypo in it, using a thermometer as stirrer. Note the temperature at which dew is first

deposited. Allow the calorimeter to warm up and note the temperature at which the dew evaporates. The cooling should be stopped as soon as the slightest signs of dew have appeared on the vessel.

WEIGHT OF AIR

9. Construct the balance described in Chapter III and measure the weight of a litre of air with it.

10. It would also be of interest to measure the weight of a litre of town gas.

BUOYANCY IN FLUIDS

11. Construct the hot air balloon described in Chapter IV, p. 47.

12. Construct the narrow cell mentioned in Chapter V, p. 56 and shown in Plate V.1, and demonstrate the production of convection currents in water. The cell is constructed by squeezing a piece of thick rubber tubing (held in the shape of a U by a piece of wire inside it) between two glass plates. The squeezing is accomplished by placing the

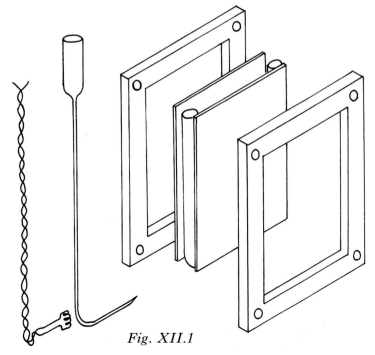

Fig. XII.1

glass plates between two face plates of wood, which are held together by
means of screws and nuts. When finished it should be about 1 cm wide.
The size of the opening will depend upon the purpose for which it is to
be used. The convection currents make a quite magnificent projection if
a simple lantern giving the image the right way up is used (see below).
For such a purpose the opening will need to be the size of an old-
fashioned lantern plate (8 cm (3¼ in) square).

The heater is made by winding a few coils of florists' iron wire round
a stout knitting needle as former. The ends are soldered to a piece of flex
– or simply twisted on if the apparatus is required only temporarily.

A solution of potassium permanganate, which is used to colour the
water, is more dense than pure water and will settle to the bottom of the
cell if let in very slowly, so that it does not stir up the contents of the
cell. It should be only a very pale pink or the convection currents may not
rise more than a short distance.

The heater can be operated by means of a dry cell from a torch battery.
The heated coloured liquid rises in a column, reminiscent of a cumulus
cloud, and a tendency for the vortex ring type of circulation can be seen.
If the heating is continued the hot coloured water will spread out under
the water surface like the anvil of a thunderstorm. When the cooled

Fig. XII.2

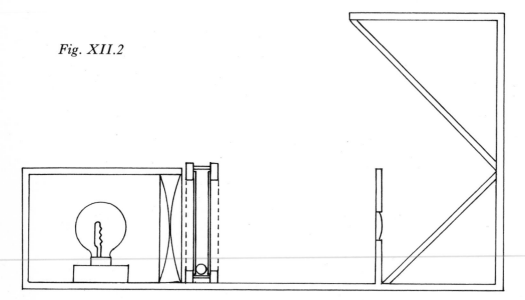

coloured water descends it does so in the form of bulbous protuberances, very like cumulus mammatus.

Other things which might be tried will suggest themselves. For example, try pouring a layer of hot water on the surface of colder water in the cell, by means of a second fine glass tube, so as to create an inversion, and see if there is a tendency for the rising cumulus to spread out underneath it. Does the inversion become marked by a layer of stratus? Can the heating be increased by adding another cell to produce a convection current which can penetrate it?

A simple lantern giving the image the right way up can be made as shown in Fig. XII.2. A simple lens of 10 to 15 cm focal length makes a surprisingly good objective; a motor car head lamp run from a battery or transformer provides the light; the erecting device consists of two pieces of ordinary plate glass mirror, which can easily be purchased from a glazier, arranged so as to project the image back over the top of the lantern. The only comparatively expensive item is the condensing lens, which has to be large enough to cover the field required, which should be about the size of the old-fashioned lantern plate.

13. Blow vortex rings in smoke as described in Chapter V, p. 66. Send them vertically to imitate the circulation in a simple cumulus cloud.

14. Observe cumulus clouds on a fine day and try to confirm that the circulation in them resembles that in the vortex ring projected upwards.

15. Measure the temperature of the surface of the ground on bare soil, by inserting the bulb of a thermometer just into the soil. Do the same on grass and see if there is any significant difference.

16. In fine summer weather, when the early mornings are overcast, use a minimum thermometer to find when the temperature at ground level starts to rise. Does this occur before the clouds have 'burnt off'?

17. Do the same inside a closed greenhouse, which traps radiation and enhances the effect.

18. Compare the rate of rise in temperature just before and just after the clouds have dispersed, to obtain an idea of how much radiation from the sun the clouds must have reflected.

19. Compare these values with the rate of fall in temperature on a still clear night after sunset.

THEORY OF THE FORMATION OF CUMULUS CLOUDS

20. Measure the height of the cloud base on a fine day, when cumulus

clouds are present and carried forward by a wind. Do this by finding how long it takes for the cloud shadow to travel a measured distance – say 100 metres – by timing the shadow between two marks at a known distance apart and parallel to the direction of the wind. Then set up a simple sighting arrangement to measure the angular velocity of the clouds overhead. For example, set up a pole vertically in the ground and fasten a crosspiece to the top on which are fixed two markers (say two nails driven partly into it). Turn the crosspiece parallel to the wind. Arrange a bottom sight on the pole so that the markers subtend a known angle (about 20°) at it. Time a cloud overhead between the marks. In that time we know from the shadow measurements how far the cloud will actually have travelled. Then, by means of a scale drawing (or by calculation using trigonometry), the height at which the cloud must be can be determined.

21. At the same time as No. 20 measure the dew point and the temperature of the surface air. Allowing 1° C per hundred metres ascent as the rate of cooling of rising air, would you consider that your measurements confirm or are in disagreement with the theory of the formation of cumulus clouds which has been given?

22. Keep a careful look-out for unusual examples of clouds and try to find an explanation for what is happening.

23. Photograph any such examples so that the photographs may serve as a record and a basis for discussion later.

24. Note any occurrence of alto-cumulus castellatus clouds. Keep a record and note on how many occasions they were followed by thunderstorms within 24 hours, within 48 hours, etc. How reliable an indication of the advent of thundery weather do they offer?

25. When next a halo appears round the sun or moon measure its angle. In order to be prepared when one does appear, design the method of procedure beforehand.

26. How reliable is a halo as an indication of approaching bad weather?

THE WEATHER MAP

27. Make a record of the direction of the wind. Obtain the current weather maps from the Meteorological Office and correlate your readings with the isobars on the map.

28. Make a similar record of the type of clouds to be seen and correlate it also with the isobars on the weather map.

29. Similarly correlate measurements of temperature with the weather map.

30. Make a rain gauge and correlate rainfall with the weather map.

THUNDERSTORMS

31. Note the direction of lightning flashes and measure the time between flash and thunder. Hence obtain the distance of the flash and plot the path of the storm on a map.

32. Make an anemometer (or wind speed measurer). A very useful arrangement is to mount a light propeller on a small permanent magnet electric motor, such as can be bought to operate certain toys. This will act as a dynamo and, if connected to a milliammeter by a length of flex (inserting a resistance if necessary), a reading of the wind speed may be made in a convenient position nearby. Instead of a propeller the usual anemometer cups can be mounted on arms attached to the motor. The propeller will need to be carefully balanced, since it spins about a horizontal axis, and it must be mounted on a weather vane, so as always to point into the wind. The usual anemometer arrangement has the advantage of making these precautions unnecessary. Either instrument can be calibrated by holding it aloft in a motor car driven at a series of speeds. The instrument must be held clear of the distorted air stream near the car.

33. Measure the wind speed and direction as a thunderstorm approaches. Do the measurements support the views of what happens in a thunderstorm which have been given in Chapter X?

34. Measure the temperature of the air during the passage of a thunderstorm. Are these observations also in agreement with the theory?

35. Collect and weigh hailstones. Estimate their diameter. Keep a record.

36. Using the density of the hailstones obtained above, make actual size models in wood of the same density. Find their terminal velocity of fall by timing their descents when dropped from two high windows at different heights above the ground. This velocity will be of the same order as that of the up-currents which maintained the hailstones in the storm during their formation.[1]

---

[1] With large hailstones the value of the terminal velocity will not be obtained very accurately unless considerable light is available. Nevertheless, the experiment will set a lower limit for the velocity of the up-currents.

HALOES AND RAINBOWS

37. Make the 60° hollow prism described in Chapter X. Fill it with wax to make the join – not too much should be applied – and make sure that it is thoroughly melted by the flame so that it runs together and makes a good seal. Allow the joint to cool a little, but not sufficiently for it to get brittle, and then seal on the next plate in the same way. When the sides of the prism have been joined together the top and bottom may then be sealed on similarly. A small aperture should be left at the top for filling. Fill the prism with boiled water. With tap water bubbles are apt to form on the walls of the vessel. The opening can then finally be closed with a last blob of sealing wax. A photograph of a water prism made with sealing wax joints is reproduced in Plate XI.1.

The colour filter for cloud photography described in No. 1 above, is made in the same way.

38. Measure the radius of the halo produced by the water prism. Compare it with the radius of an actual halo measured in No. 25. The value obtained with water will probably be slightly greater than that obtained in an actual halo because of the difference between the refractive indices of water and ice.

39. Note the type of cloud in which haloes appear.

40. When a rainbow is seen, take the opportunity of measuring the angle it subtends at the eye. A complete circle is very rarely indeed to be seen, and then only from the air. It will be necessary, as for the halo, to work out beforehand the procedure to be followed.

41. Do the same for the secondary bow, when one is visible.

42. Note the colours in the rainbow carefully. In spite of what is said in some books, rainbows are not all alike. See if any differences are detectable. To do so will entail keeping a very careful record. Coloured photographic records can be very helpful.

43. It has been suggested that rainbows are not necessarily circular. Could a test of this suggestion be devised?

44. Draw a circle of radius about 10 cm on a piece of drawing paper and, taking the refractive index of water to be 4/3, repeat Descartes' investigation of the rainbow by drawing a series of rays going into a water drop, being reflected off the back surface and then emerging again. Measure the angle of deviation and plot a graph of deviation against the angle of incidence. Does the graph show a minimum deviation and if so what is its value?

45. Repeat for the secondary bow by drawing rays suffering two internal reflections instead of one.

46. Record the colours in the supernumerary rainbows to be seen occasionally within the primary bow.

47. Make a lycopodium halo. To prepare the diffracting screen of lycopodium dust first make a narrow glass cell to hold the lycopodium by sticking celluloid strips (about a millimetre thick) round the edges of a 5 by 5 cm lantern slide cover glass, using 'Durofix' or similar cement. Allow it to stand under light pressure for 24 hours, to set.

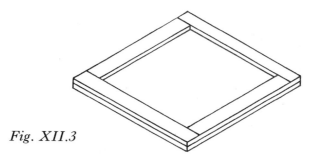

Fig. XII.3

Take a few cubic centimetres of warm water and dissolve as much gelatine – such as is used for table jellies – in it as it takes to give a gel which will set in a minute or two when cooled. While still hot add a very small pinch of lycopodium powder and shake well, so as to colour the mixture a very pale yellow. Pour some into the cell just prepared and close carefully with a cover glass, being careful not to entrap air bubbles. Allow to set. It is a good idea to prepare several cells and gradually to increase the proportion of the lycopodium powder. Start with a very weak suspension. Pour one cell. Then add a little more powder to the remainder and pour the next, and so on. In this way it is then possible to choose the best screen for the purpose in hand. The halo can be seen simply by viewing an illuminated pinhole through the screen held close to the eye. It is best to try the experiment in a darkened room. Details of how the halo can be projected are given in Chapter XI.

If it is desired to take a coloured photograph of the rings, a telephoto lens (one of focal length 135 mm is suitable with a 35 mm camera) should be used to photograph an illuminated pinhole about five metres (fifteen feet) away, with the lycopodium screen held in front of the lens. A convenient method for finding the correct exposure is to take some pre-

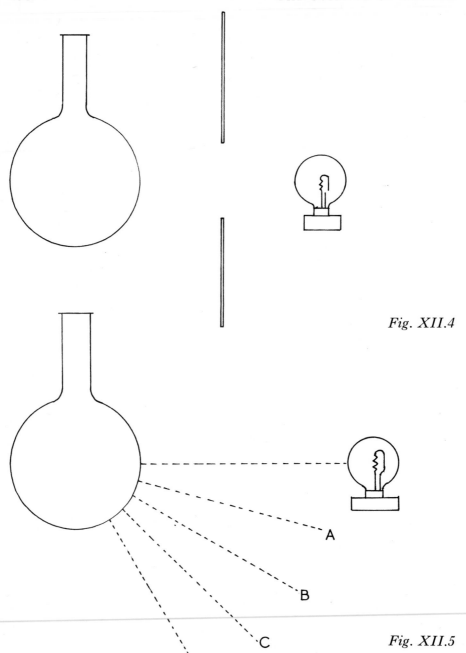

Fig. XII.4

Fig. XII.5

liminary photographs in black and white, which can be developed on the spot and used to get the exposure required – allowing, of course, for the differing speeds of the black and white, and the colour film.

48. Place a small lamp behind a circular aperture in a piece of white card or stiff paper, and hold a spherical flask of water in the beam, (Fig. XII.4). Note the reflected beam where it strikes the card and use it to illustrate the formation of the rainbow. Estimate the angles and use the value to forecast the angle of the rainbow. Does the value obtained agree with the angle of the rainbow actually measured? If there is any difference, to what is it due? How can it be minimised?

49. Set up the spherical flask and small lamp as in Fig. XII.5, and note the position of the images to be seen reflected in the flask. Describe carefully what happens to them as the eye is moved round from the position marked A, near to the lamp, through B and C, to D. Develop a method for distinguishing between the image caused by reflection at the front surface and that formed by reflection at the back. Also ascertain

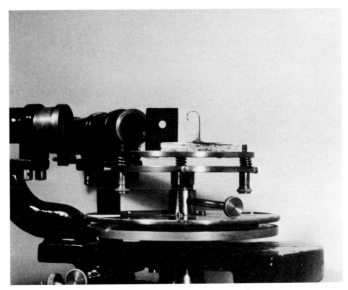

Plate XII.2   Measuring the angle of the rainbow by means of a spectrometer. A small drop of water is suspended from the glass hook coated with beeswax. The angles of both primary and secondary rainbows can be measured by means of the telescope.

which of the latter was caused by a single reflection and which by two such reflections.

50. If a spectrometer is available measure the angle of the rainbow as follows. Draw out a piece of narrow glass tubing to a very fine diameter. Cut off a piece and bend it into the form of a hook. Mount it vertically by means of sealing wax on a piece of plywood, which can stand on the table of the spectrometer as in Plate XII.2. The end of the hook should be on a level with the centre of the lens of the collimator. Warm the glass hook slightly and touch it on a piece of beeswax so as to coat it with a film of wax to make it water-repellent. With patience a small drop of water can be persuaded to hang on to the end of the hook, and if this has been made water-repellent with beeswax the drop will assume a spherical form. The light from the collimator which would miss the drop, should be screened off by a piece of card with a small hole in it level with the drop. The angles of both primary and secondary bows can be measured in this way. The general illumination within the primary and outside the secondary bow can also be detected.

51. Use a garden syringe to create a spray of fine drops of water in sunshine. Observe the rainbow which can be produced. In which direction must the spray be directed? Is the general illumination within the bow visible?

52. Scatter some small polystyrene spheres over a piece of black paper and mount a small electric lamp with a small filament (such as a torch bulb) some way above them. Observe the rainbows which are produced. Why are there two of them?

Index

Figures appearing in bold refer to the colour plates.

QC
921.5
.T7
1970

QC
921.5
.T7

1970